GHOST
NATION

ABOUT THE AUTHOR

Chris Horton is a reporter who has covered cross-strait politics, domestic politics, the economy, culture and breaking news in Taiwan for *The New York Times*, Bloomberg News, *The Atlantic*, *The Guardian*, *Financial Times* and *Nikkei Asia*. He has lived in Taiwan since 2015, and in China and Hong Kong for the decade preceding that. *Ghost Nation* is his first book.

CHRIS HORTON

GHOST NATION

The Story of Taiwan
and Its Struggle for Survival

MACMILLAN

First published 2025 by Macmillan
an imprint of Pan Macmillan
The Smithson, 6 Briset Street, London EC1M 5NR
EU representative: Macmillan Publishers Ireland Ltd, 1st Floor,
The Liffey Trust Centre, 117–126 Sheriff Street Upper,
Dublin 1 D01 YC43
Associated companies throughout the world

ISBN 978-1-0350-3402-4 HB
ISBN 978-1-0350-3403-1 TPB

Copyright © Chris Horton 2025

The right of Chris Horton to be identified as the
author of this work has been asserted in accordance
with the Copyright, Designs and Patents Act 1988.

All rights reserved. No part of this publication may be reproduced,
stored in a retrieval system, or transmitted, in any form, or by any means
(including, without limitation, electronic, mechanical, photocopying, recording
or otherwise) without the prior written permission of the publisher.

Pan Macmillan does not have any control over, or any responsibility for,
any author or third-party websites (including, without limitation, URLs,
emails and QR codes) referred to in or on this book.

1 3 5 7 9 8 6 4 2

A CIP catalogue record for this book is available from the British Library.

Map artwork by ML Design Ltd

Typeset in Perpetua Std by Six Red Marbles UK, Thetford, Norfolk
Printed and bound in the UK using 100% Renewable Electricity by CPI Group (UK) Ltd

This book is sold subject to the condition that it shall not, by way of
trade or otherwise, be lent, hired out, or otherwise circulated without
the publisher's prior consent in any form of binding or cover other than
that in which it is published and without a similar condition including this
condition being imposed on the subsequent purchaser. The publisher does not
authorize the use or reproduction of any part of this book in any manner
for the purpose of training artificial intelligence technologies or systems.
The publisher expressly reserves this book from the Text and Data Mining
exception in accordance with Article 4(3) of the European Union
Digital Single Market Directive 2019/790.

Visit **www.panmacmillan.com** to read more about
all our books and to buy them.

To the Taiwanese people, and my parents, Jim and Valerie

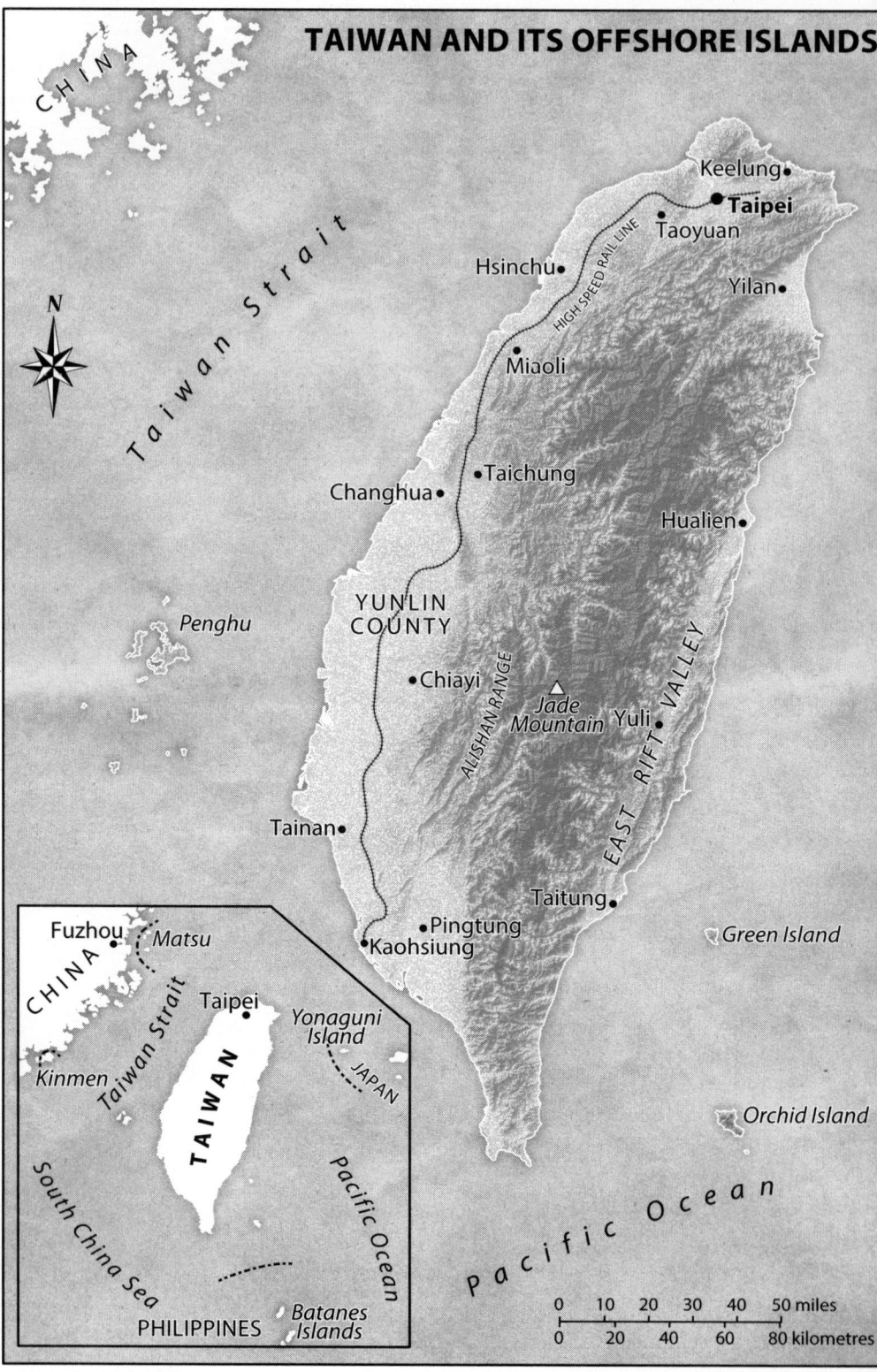

CONTENTS

INTRODUCTION Dark Clouds
1

ONE Appeasing the Afterworld
13

TWO Ilha Formosa and its Colonizers
27

THREE In the Shadows of Empire
43

FOUR Shifting Winds, Deadly Storm
55

FIVE White Terror in Free China
83

SIX Death of a Dynasty
113

SEVEN The Three Immortals
135

EIGHT The Democracy Experiment
157

NINE Taiwan in the Balance
189

TEN Scrubbing Taiwan's Sovereignty
215

ELEVEN The Indispensable Island
255

EPILOGUE Eye of the Typhoon
273

ACKNOWLEDGEMENTS
283

LIST OF ACRONYMS
287

NOTES
289

FURTHER READING
309

INDEX
311

INTRODUCTION

Dark Clouds

Next time you're feeling overwhelmed at work, spare a thought for Lai Ching-te.

As Taiwan's president since May 2024, Lai has a lot on his plate. In 2025, Taiwan joined the ranks of the world's superaged societies, where at least 20 per cent of the population is aged sixty-five or older.[1] That trend is set to continue well into the future – the country's fertility rate is among the lowest in the world. Simultaneously, Taiwan is suffering from a nationwide labour shortage across multiple sectors, including the government and military.[2] Although opening the country up to immigration beyond certain preferred demographics such as white-collar workers, professional athletes and digital nomads seems like an easy fix, it is probably more than a largely conservative Taiwanese electorate can stomach.

As Taiwanese society adjusts to cope with the growing burden of caring for its elders, its youth are increasingly disillusioned by their prospects. For the past twenty-five years, the cost of living has steadily increased, while wage growth has lagged behind. For most young Taiwanese, home ownership is an elusive dream. The discontent among many young voters is aimed at Lai's party, the Democratic Progressive Party (DPP), which has held the presidency since 2016, during which time it has gone from being the opposition to the establishment.

Taiwan's leading semiconductor firm and one of the world's most valuable companies, TSMC, is building massive multi-billion-dollar chip factories that will require a large and steady supply of electricity, but the power grid is under growing strain. Nuclear power seems like an obvious solution to many outsiders, but is a politically fraught subject. The semiconductor sector's immense water requirements are also threatened by climate change, with droughts becoming more frequent and severe in recent years.

And then there's China.

The Chinese Communist Party (CCP) claims Taiwan as its rightful territory, even though it has never ruled it. Stern-faced CCP officials insist that democratic and independent Taiwan belongs to the authoritarian People's Republic of China (PRC), but the facts speak for themselves.

Declaring Taiwan to be part of China is like saying North America belongs to Europe. Both sets share historical links through migration and colonialism, yet retain fundamentally distinct identities. Just as the first people to populate North America were not European, Taiwan's original inhabitants are not Chinese, arriving from the Asian mainland before Chinese culture existed. In all of human history, there have only been four brief years during which a China-based government ruled all of Taiwan. That was the Republic of China (ROC) government in Nanjing, which was more focused on looting than administering Taiwan between 1945 and 1949. In 1947, less than two years into being part of the ROC, the Taiwanese people rose up violently in every major city, and were slaughtered in their tens of thousands. Hardly the stuff of which historical claims are made.

Before the ROC arrived, it was Imperial Japan that had conquered and controlled Taiwan after previous empires from Europe and the Asian mainland failed to fully subjugate it. Japan ruled

Taiwan from 1895 to 1945, but it took two decades of bloody 'pacification' campaigns for Tokyo to effectively administer the entire island. While Taiwan had previously been colonized piecemeal by European and Asian colonial aspirants, it was Japanese colonialism that first unified all of Taiwan under one government. The Japanese colonial project profoundly shaped the agriculture, industry, education, cuisine and culture of Taiwan. Both the later ROC military dictatorship and the CCP party-state that seeks to annex it today have done their best to downplay Japan's massive influence on Taiwanese identity. This makes good sense, as this transformative Japanese influence undercuts their assertion that the Taiwanese are Chinese.

It wasn't until 1943 that the ROC's Chinese Nationalist Party – better known as the Kuomintang, or KMT – or its successor in China, the CCP, made claims on Taiwan. Eight decades later, the CCP is now closer than ever to being able to subsume Taiwan, a goal that has become a top priority under the current leader of China, Xi Jinping. Following Xi's rise to power in 2012, the CCP has intensified its efforts to undermine Taiwan's sovereignty and hard-earned democracy. Since 2020, every year has seen a number of unpleasant new normals – the result of relentless but incremental Chinese 'salami-slicing' tactics. For Taiwanese people living on the east and west coasts, the roar of fighter jets scrambling to intercept Chinese warplanes just beyond Taiwanese airspace is now part of daily life. Only in Taiwan's north, with Taipei at its centre, do people live without these regular reminders of the CCP's sinister intent. For now, at least.

For those living on Taiwan's outlying islands, the sight of Chinese fishing boats and dredges stealing fish, squid and sand from their waters has become commonplace in recent years. Well before China-connected boats were found to be involved in the severing of undersea cables in Europe in 2024, the same thing played out twice in Taiwan's Matsu archipelago in early 2023. Since then,

China-related ships have been accused of suspected cable-cuttings in waters near Taiwan and Europe.

Beijing is ramping up its aggression in other domains as well. Taiwan is constantly subjected to cyberattacks from China, which are becoming both more frequent and more sophisticated. These attacks target the government and military as well as companies and universities. Beyond technological intrusions, Taiwan faces a barrage of Chinese disinformation through both traditional and new media platforms. The main messages of China's cognitive warfare on the Taiwanese people are:

- You are Chinese and China's rise is unstoppable – come join the winning team.
- The Americans are using you to split and contain China and will eventually abandon you in your moment of need.
- If you surrender before any shots are fired, we'll let you keep all the freedoms you have now – all who resist will be destroyed.

Moreover, Chinese influence is increasingly noticeable in Taiwan's legislature. In early 2024, the two main opposition parties formed a majority coalition that has consistently promoted an agenda that aligns with many of Beijing's goals. This agenda promotes a Chinese identity, seeks to defund the government and aims to weaken the constitutional court. At such a crucial juncture, when national unity is imperative to counter the urgent threat posed by China, the proxies of the CCP in Taiwan's legislature are serving as a wrecking ball.

Lai Ching-te was only two years old when his coal miner father passed away, leaving him as one of six children in a struggling family. Lai would eventually work his way out of poverty to become a physician and, later, a legislator, mayor, premier and vice president. He has often credited his father for having left him very little, as growing up poor compelled him to work hard and overcome the challenges of his early life.[3] Few challenges, however, are as formidable as defending Taiwan's sovereignty from an increasingly bellicose China. During his campaign for the presidency and after his election, Lai made multiple efforts to re-open informal communication between Taipei and Beijing, which had been abruptly cut off by Xi in 2016. Beijing remains steadfast in its refusal to engage in dialogue with Lai's government unless he publicly endorses the notion that Taiwan belongs to China. For Lai, whose public persona is very much being a son of Taiwan, that is not going to happen.

Within living memory, the Taiwanese people have endured levels of repression and fear of state retribution similar to those faced by Chinese citizens today. This is one of many reasons that the Taiwanese population overwhelmingly reject any union with China. For many Taiwanese, any animosity towards China is aimed primarily at the CCP, not the Chinese people. The Taiwanese are acutely aware of the long journey to freedom, having once been even more isolated and politically indoctrinated than China is today. In the course of my reporting in Taiwan, both older and younger Taiwanese have expressed sympathy for their Chinese neighbours. Whether from first-hand experience or through stories told by elder relatives, the Taiwanese see echoes of their recent dark past in Xi Jinping's modern China.

Meanwhile, Xi's preparations for a war of conquest in Taiwan continue apace. In late 2024, China's People's Liberation Army (PLA) conducted its third large-scale joint force drills around Taiwan that year. This display was the biggest show of naval force by China against its democratic neighbour since the mid-1990s, when Taiwan was preparing for its first presidential election. During that tense period, a potential Chinese attack on Taiwan was averted by the deployment of two carrier groups from the US Seventh Fleet on the orders of President Bill Clinton. But the spectre of war looms once again over Taiwan, and China's military has dramatically modernized, upgrading in both size and capability.

Currently, the PLA boasts the world's biggest navy, which includes two aircraft carriers and a third on the way. It continues to churn out amphibious attack ships while also incorporating the world's largest civilian ferries into its war drills. Although the PLA Rocket Force is untested in actual combat, it is considered world-class. In any areas where China lags behind the US, it is quickly closing the gap – including its nuclear arsenal, which is being built up rapidly and in extreme secrecy. As Beijing intensifies pressure on Taiwan across various domains, Chinese military planners are clearly preparing for more than a simple confrontation against its smaller neighbour. Using the vast, sparsely populated regions in western China for exercises, the PLA Rocket Force has conducted long-distance missile attack simulations on targets that resemble American and Japanese warships. One of its missiles, the Dongfeng-26, is nicknamed the Guam Express, after the US Pacific territory that would be crucial to American efforts to help Taiwan.[4]

China has threatened retribution against Japan, the Philippines and Australia should they provide any assistance to Taiwan or the US in any conflict. In 2023, the Chinese ambassador to the Philippines warned Manila against granting the US military access to northern bases near Taiwan – if it cares about the 150,000 overseas Filipino workers in Taiwan. China's main target beyond Taiwan is, of course, the US, and conventional military hardware is not the only weapon at its disposal. The Chinese hacking group Volt Typhoon has infiltrated critical American infrastructure – including ports, water treatment facilities and oil- and gas-processing units. Their targets include Guam and Hawaii, which are vital for US support of Taiwan. Creating chaos within the US economy – and the general public – before or during an attack on Taiwan could buy China crucial time to establish a foothold, making a takeover inevitable before US forces even have the chance to respond.

In recent years, closer ties between China, Russia and North Korea have also raised the possibility of coordinated actions between the three authoritarian countries to overwhelm the US military. One possible scenario involving simultaneous actions would involve Russia targeting NATO members or Japan; North Korea attacking or provoking South Korea; and China launching an invasion of Taiwan. Such coordinated efforts would significantly impair Washington's ability to respond in support of Taiwan (and indeed its other treaty allies).

If Xi chooses to launch an elective war on Taiwan, it would likely escalate into a regional war at the very least. War games conducted in Washington, DC and Taipei consistently predict massive Taiwanese, Chinese and American loss of life, irrespective of Chinese success or failure. Such a conflict could also involve nuclear brinksmanship. Economically, the repercussions would be global. Chinese, American and international economies would suffer, with most estimates envisioning damage far beyond the fallout of the Covid-19 pandemic. In 2024, Bloomberg estimated a Chinese war against Taiwan would deliver a US$10 trillion hit to the global economy — that's 10 per cent of the planet's GDP.[5]

But why would Xi Jinping want to wreak such havoc? The PRC is the size of the US and boasts the world's second-largest economy. It is already a global power. Why would he want to risk his country's recent gains for a mostly mountainous island the size of Belgium that it has never controlled? There are two primary reasons, both centred around power.

First, annexing Taiwan and administering it as PRC territory would significantly alter Beijing's strategic landscape, giving China a Pacific coastline for the first time. This would dramatically expand

China's power-projection capabilities. Currently, China's coastline is restricted by what is known as the First Island Chain: a series of democratic countries including South Korea, Japan, Taiwan and the Philippines. Taiwan is the closest link in this chain to China's coastline, and is the only one lacking a mutual defence treaty with the US, making it the most vulnerable. If China were to absorb Taiwan, it would instantly become a Pacific power. Chinese submarines, which are relatively easy to track within the shallow waters inside the First Island Chain, would suddenly be able to drop into the depths right off Taiwan's Pacific coast. With its location at the intersection of the Pacific, the Sea of Japan and the South China Sea, Taiwan would enable China to control the flow of global commerce through the world's busiest waterways. Beijing would be able to turn the tap of global trade on and off at will, for any or all countries. If the PRC were to swallow Taiwan, it would also result in Chinese military bases popping up on the doorsteps of Japan and the Philippines. Both nations currently administer islands and other maritime features also claimed by Beijing, and Taiwan's annexation would significantly increase the Chinese military's capacity to threaten these two US allies, as well as extend its influence further southwards into maritime Southeast Asia.

Second, conquering Taiwan would be a massive morale and propaganda boost for Beijing, both at home and abroad. Domestically, it would elevate Xi Jinping's status above that of Mao Zedong as Communist China's most powerful leader. It would also eliminate Taiwan's democracy, which, through the struggles and sacrifices of its people, replaced the Chinese authoritarian regime that arrived in the 1940s. For those inside and outside of the PRC advocating for a free and democratic China, the fall of Taiwan would be a devastating setback. Globally, the political windfall that a successful annexation of Taiwan would give China is difficult to overstate. If the US was unable or unwilling to mobilize in support of Taiwan, its Asian allies and partners would have to adjust accordingly – doubly so if the US military lost to China in a conflict.

Either way, the CCP would emerge as the dominant force in the world's most populous and dynamic region, amplifying its global clout. Many Asian countries might shift closer to China, making concessions to maintain favourable relations with Beijing. Others, such as South Korea, Japan or Australia, might consider developing their own nuclear deterrents.* Or they may simply move closer to China too. Regardless of the outcome, democracy would likely recede in Asia, while China and its allies – Russia and North Korea – would be emboldened. Those in Taiwan who survive a Chinese invasion would likely be subject to martial law and cut off from the rest of the world, just as Beijing has done to East Turkestan (which it calls Xinjiang) and Tibet (now being rebranded as Xizang). These areas have seen the establishment of concentration camps and efforts aimed at forced assimilation into the PRC.

If Beijing's handling of Uyghurs and Tibetans is a guide, a successful Chinese invasion would force the Taiwanese people, for the second time in less than a century, to relinquish their history and identity or face severe consequences, including torture, imprisonment and even execution. This echoes the Taiwanese people's past experience with Chinese martial law from 1949 to 1987 – and mirrors the current struggles of Hong Kong's democratic movement today. In response, a robust diasporic resistance movement of Taiwanese abroad would likely emerge, fighting to preserve their cultural legacy and raise awareness of human rights violations in their homeland. Such a resistance movement would be the target of a transnational repression campaign by Beijing, likely including sanctions, bounties, kidnappings and potentially even assassinations.

---

* It is worth noting that Taiwan abandoned its nuclear weapons programme in the 1980s, and Ukraine gave up its nuclear arsenal in the 1990s, both largely predicated on American security promises.

Meanwhile, just as in Manchuria, East Turkestan, Tibet, Southern Mongolia and Hong Kong, Han Chinese immigrants would replace local populations through mass migration. But the Taiwanese people, exiles and diaspora abroad would not suffer in isolation. The loss of Taiwan and the disruption of the US-backed post-Cold War security order would have global repercussions. A conflict would also devastate Taiwan's world-leading semiconductor industry, as China increasingly dominates less advanced chip production. Developed economies, including the US, would suddenly find themselves desperate for semiconductors, with China holding most of the chips. China would have significant leverage over the global economy and geopolitical landscape.

The return of Donald Trump to the White House in January 2025 has added another degree of uncertainty to Lai's job. Between his first and second term as President, Trump was critical of Taiwan, suggesting it needed to pay more for American protection and falsely claiming Taiwan had stolen the US's semiconductor industry. Just days before Trump's re-inauguration, Foreign Minister Lin Chia-lung said, 'In response to President-elect Trump's call for Taiwan to demonstrate a stronger commitment to self-defence, we are of course ready to increase our defence budget.'[6] But Lin emphasized that Taiwan is not only asking for help, it is also offering it. Beyond semiconductors, Taiwan could serve as a manufacturing hub for drones and other military equipment that the US and other countries are struggling to produce at scale, he said. It seems that the Lai administration's approach to the second Trump era is to position itself as a proactive partner, rather than a protectorate.

If the US, Japan and other countries are to work with Taiwan to hold the line against Chinese expansionism, understanding Taiwan's internal political dynamics will be crucial. This can only be done by understanding who the Taiwanese people are, and the history that has shaped their country. Despite growing global awareness of Taiwan and its importance, especially since 2020, it remains poorly understood. For countries whose stability and prosperity depend on Taiwan's continued sovereignty, it is vital for policymakers, journalists and voters to be able to differentiate between truth and lies about Taiwan.

This book, the culmination of hundreds of interviews with Taiwanese of diverse backgrounds over a decade of reporting in Taiwan, seeks to dispel the carefully crafted disinformation sowed by Beijing and its proxies about Taiwan. It underscores the weakness of China's claims, and follows the Taiwanese people's intergenerational struggle for the independent and sovereign country they enjoy today. Ultimately, it highlights how Taiwan is the fulcrum upon which the world's balance of power now rests, and a partner that fellow democracies abandon at their peril.

As President Lai emphasized in his 2025 New Year address, 'The more secure Taiwan is, the more secure the world is. The more resilient Taiwan is, the sounder the defence of global democracy will be.' For democratic governments worldwide, most of which maintain unofficial relations with Taipei, the importance of Taiwan to their national interests is clearer than ever. Working with Taiwan to staunch the spread of authoritarianism, however, requires understanding Taiwan and its people.

CHAPTER 1

Appeasing the Afterworld

On a scorching hot summer morning in Taipei, the city pulsing with the energy of rush-hour traffic and the relentless hum of urban life, a profound silence seemed to envelop the entrance to Taiwan's presidential office building. If you happened to be walking past on the morning of 1 August 2016, you might have caught a glimpse of an ancient Paiwan ceremony before moving on your way – loitering in front of the building or taking photographs is not permitted, as any of the white-helmeted armed guards out front will tell you.

Before the grand brick edifice of the presidential building, its central tower rising high above its royal palm-lined facade, a scene of both spiritual and political importance unfolded. Dozens of men and women had assembled, all donning the colourful traditional garments of Taiwan's original Indigenous peoples. They stood in four columns and observed as five male Paiwan elders hunched over in a circle and chanted. As a white wisp of smoke floated upwards from the centre of the circle, the men rose and used a wooden pole to raise a smoking bundle of millet stalks towards the sky. They were inviting ancestral spirits to the morning's event: the first apology by any government of Taiwan to its original inhabitants after 392 years of colonization.

Once the invitation had been sent, the handful of Amis delegates, led by two men in striking red robes and sunglasses, were the first to walk up the carpet into the presidential building. There, they were greeted by President Tsai Ing-wen, who shook their hands before they proceeded inside for the ceremony. The Amis were followed by a procession of their Indigenous neighbours on the island, each announcing their presence with a collective whoop and the loud blaring of air horns. Each group's clothing and appearance varied significantly, with different materials, styles, colours and adornments such as leather, boar tusks or feathers. First were the other fifteen nationally recognized ethnicities: the Atayal, Paiwan, Bunun, Puyuma, Rukai, Tsou, Saisiyat, Tao, Thao, Kavalan, Truku, Sakizaya, Seediq, Hla'alua and Kanakanavu. They were followed by representatives of the Indigenous *Pingpu* plains peoples who still seek government recognition: the Siraya, Ketagalan, Taokas, Pazeh, Taivoan, Papora, Makatao, Hoanya, Kahabu and Babuza tribes. This mosaic of peoples, each draped in their unique attire, reflected the rich tapestry of Taiwan's Indigenous heritage, which stretches back millennia before the arrival of Dutch and Chinese 'double colonization' in the seventeenth century.

Within the walls of the presidential building, Hudus Haitang, an elderly spiritual medium of the Bunun tribe, took to the stage to lead a rite intended to ensure both the success of the ceremony and improvements to government Indigenous policies. Clad in a black shirt hemmed with colourful embroidery and adorned with silver, the medium welcomed Tsai to the stage. Her hair in her trademark bob, the recently inaugurated president wore a grey suit jacket, its lapels featuring the black snake-inspired motifs favoured by the Paiwan – Tsai herself being of partial Paiwan descent. Together, they performed the rite, sprinkling wine over reeds in a rattan basket. Hudus chanted a prayer as she shook the reeds and asked the spirits to guide Tsai towards wise decisions during her presidency. The rite's objective, the audience was told, was to introduce Tsai to the spirit realm so that she could receive its blessing and protection.

Highlighting the influence of Western colonialism on Taiwan's Indigenous peoples, six Christian ministers from different tribes took to the stage. While some Indigenous people still practise the animism of their ancestors, well over half are Christian – Indigenous villages in the country's rugged mountainous regions tend to have two churches, one Catholic and one Presbyterian. Their arms raised and outstretched towards the audience, the ministers each recited a Christian prayer in their native tongue before they finished together in Mandarin. Vice President Chen Chien-jen, a devout Catholic, lowered his head as he sat listening in the front row.

With the ancestral spirits and the Christian God now all invited to bear witness, Tsai returned to the stage with Syapen Nganaen, a male elder of the Tao tribe. The Tao are expert boat makers and primarily live on Orchid Island in the Philippine Sea, around 60 kilometres off Taiwan's south-eastern coast. Syapen Nganaen was the event's designated representative of Indigenous Taiwanese peoples. He wore a light pink headband, a grey-and-white-striped vest worn over a white T-shirt and a sumo-like thong bottom. The two bowed before each other and then shook hands at centre stage before returning to their respective lecterns on either side. Tsai spoke first.[1] She noted that the ceremony was being held on the twenty-second anniversary of the inclusion of Indigenous people in the national constitution, in which their official designation was changed from *shanbao* (山胞 – something like 'mountain kin') to *yuanzhumin* (原住民 – 'Indigenous people').

'This correction not only did away with a discriminatory term, but also highlighted the status of Indigenous peoples as Taiwan's original owners,' Tsai said. 'From this starting point, we are taking another step forward today. To all Indigenous peoples of Taiwan: on behalf of the government, I express to you our deepest apology. For the four centuries of pain and mistreatment you have endured, I apologize to you on behalf of the government.

'The Dutch and the Koxinga clan's kingdom massacred and exploited the Pingpu ethnic group,' Tsai said of the first outsiders to seize territory in Taiwan. 'The Qing Empire presided over bloody confrontations and suppression. Colonial Japan put in place comprehensive "savage" policies. And the post-war Republic of China government undertook assimilation policies.

'For 400 years, every regime that has come to Taiwan has brutally violated the rights of Indigenous peoples through armed invasion and land seizure,' Tsai continued. 'For this, I apologize to the Indigenous peoples on behalf of the government.'

After more apologies for a litany of sins committed by previous leaders and governments, Tsai concluded and allowed Syapen Nganaen to give the Indigenous response. The Tao elder was dignified and righteous as he addressed the audience in a building that was designed by Japan, the first empire to conquer his native Orchid Island and the Tao people.

'In just a few days I'll turn eighty-one years old,' Syapen Nganaen said, speaking in the Tao language and swaying from side to side as he talked, his skinny, bare legs below the lectern contrasting with Tsai's bespoke Western-style suit. 'I was born when the Japanese ruled Taiwan – I experienced Japanese rule.'

Only days earlier, he said, he'd received a phone call asking him to come to the capital to represent the Indigenous people of Taiwan, past and present, as the government apologized for the first time to them.

'I know the government of Taiwan has had many presidents over the years, but there's never been a president who was willing to apologize to the Indigenous peoples. Only the current government of President Tsai Ing-wen has apologized to the Indigenous peoples, so I'm very happy.'

He appeared more indignant than happy, however. As he spoke, the pained look on his face intensified as he described the thirty years that Taiwan's government had clandestinely stored nuclear waste on Orchid Island. Some barrels have begun to corrode, and he and his fellow Tao were concerned that leaks could render their ancestral homeland uninhabitable, he said, shaking his hands as he spoke louder.

'We are gathered together today on this special day to accept the apology that the government has offered to the Indigenous peoples,' he said, waving his right arm and then his left and bowing multiple times as he spoke. 'Today is the beginning of reconciliation and harmony. So I hope the Taiwan government will affirm the love of God so that people will show love to each other, help each other and live in harmony.'

Syapen Nganaen thanked Tsai for facilitating the apology and expressed his hope that her government would follow through on its word.

At the end of the ceremony, Syapen Nganaen smiled for the first time when Tsai presented him with an official copy of the text of the government apology. As he looked out at the audience of his fellow Indigenous Taiwanese, he appeared to savour the moment. He then presented Tsai with a gift of millet stalks, signifying the hope of Indigenous peoples that justice and reconciliation would take root.

Seated in the front row of the audience in a sleeveless black dress complemented by an embroidered red sash and long necklaces of green, yellow and white beads was forty-two-year-old former legislator Kolas Yotaka. Just a few months earlier, she had become the presidential office's first Indigenous spokesperson. Kolas is a member of the Amis people, also known as the Pangcah, one of Taiwan's sixteen officially recognized Indigenous groups. Her family's roots are in Yuli, in the middle of the East Rift Valley between the Coastal Mountain Range and Central Mountain Range. The hundred-mile-long East Rift Valley is where the Eurasian Plate and Philippine Sea Plate meet. Verdant and fertile, it receives the mineral-rich water that streams down from both mountain ranges before it flows seawards via Hualien to the north or Taitung to the south.

Kolas's name connects her to the land that supported her family, as well as the colonization that altered the destinies of the Amis and other Indigenous Taiwanese. In the Pangcah language spoken by the Amis people, 'kolas' is a verb that describes sorting out the bad seeds when planting, so that the good seeds will grow into strong plants. It can be used by men or women, and is also the name of one of Kolas's grandmothers. The surname Yotaka was given to her grandfather during Japanese colonial rule, but following the arrival of the Kuomintang party-state from China in the 1940s, the name Yotaka was changed to Yeh. When Kolas was born in the 1970s, Taiwan was still under martial law. The KMT was imposing a Chinese identity on everyone island-wide, including the Indigenous tribes. Indigenous names were not officially recognized, which erased a major part of people's identities. Instead, the Indigenous peoples of Taiwan were forced to adopt Chinese names.

As a result, Kolas's legal name at birth was Yeh Kuan-lin. That was the name she was forced to use until 2006, when a newly democratic Taiwan passed legislation that paved the way for Indigenous citizens to change the names on their household registration to their real names. For Kolas, being able to register the name her parents gave her was a major turning point in her country's recognition of who she was – and that her people were not Chinese.

'I was happy, it felt like the natural thing to do,' she told me of her legal name change. 'Knowing that my old classmates and friends would not recognize my name, or that people might make fun of me, didn't stop me.'

For seven years prior to the legislation, she had been using her Romanized Pangcah name. Although this strengthened Kolas's connection to her Indigenous heritage, it made her life much more difficult.

'It created a lot of problems for me, but all you can do is be patient and try to educate people,' she said.

Domestic airline tickets couldn't accommodate her full name. Banks with systems designed for names written in Chinese characters turned her away rather than try to help. Nurses at hospitals wouldn't speak to her in Mandarin when they'd see her name on a patient chart. They would use their hands or body language to try to communicate with her, despite her fluency in Mandarin (she is also fluent in Taiwanese and English). But those days are long gone now.

'I love hearing people call me by my name or see them write my name because whenever they use it, it means they recognize there are different peoples in Taiwan,' she said. 'Indigenous peoples don't exist only in books and museums — we are living and breathing right next to you.'

Unlike her ancestors, Kolas has been able to see the arc of Taiwanese history begin to bend towards justice. Before entering politics in 2016, the former journalist lobbied the government for an acknowledgement of the past wrongs committed against Taiwan's original inhabitants, and a formal apology.

'I always wanted to see this happen,' she told me, saying that she and other Indigenous Taiwanese were inspired by similar efforts in other countries to confront and reconcile with their historical injustices. Among them, she said, was South Africa, where Nelson Mandela and Desmond Tutu initiated the Truth and Reconciliation Commission in 1996 to address the country's history under apartheid. The 2008 apologies by Australian Prime Minister Kevin Rudd to his country's Indigenous peoples and Canadian Prime Minister Stephen Harper to his country's First Nations inspired Tsai's apology eight years later.

'I thought Taiwan should do it too, and I think the president apologizing to Indigenous peoples was a meaningful step for Taiwan to move forward,' Kolas said.

'On a personal level, on the morning of the apology my hope was that my ancestors felt healed,' she said. 'After being discriminated against, tortured, enslaved and slaughtered, I hoped that they felt respected and dignified as human beings.'

For Kolas, the most memorable part of the ceremony was seeing Syapen Nganaen accepting the apology.

'The Tao people are the most marginalized Indigenous group in Taiwan, and the traditional clothing he was wearing, which is used on formal occasions, would be considered inappropriate in Han Chinese culture because it doesn't cover much of the body,' she said. 'I was proud of him – his face and words and body language told me that we can't stop fighting, because we haven't gotten there yet.'

During the eight years of the Tsai administration, Kolas said, much progress was made on Indigenous issues, especially compared to previous governments. It passed the Indigenous Language Act, which aimed to strengthen Indigenous language education in schools. The legal process for returning Indigenous lands was streamlined. But much remains to be done to integrate the descendants of Taiwan's original inhabitants and Taiwan's settlers. Some problems can be legislated away, but systemic social and economic marginalization will take longer to address. 60 per cent of Indigenous Taiwanese live below the poverty line. They are more likely to drop out of school, and have a life expectancy six years lower than the national average. Even in professions where Indigenous participation is widely accepted – sports, policing, the military and entertainment being most prominent – glass ceilings abound. As Kolas puts it: 'Indigenous people are players but not coaches, soldiers but not generals, teachers but not principals, singers but not producers.'

According to Taiwan government statistics from 2021, the Amis tribe is the largest Indigenous group in Taiwan, with a population of over 215,000.[2] This is more than double the number of the next two largest groups, the Paiwan and Atayal, which have approximately 104,000 and 93,000 members respectively. The contrast is stark when considering the smaller Indigenous groups at risk of extinction, such as the Hla'alua with 429 members or the Kanakanavu with only 380. Historically, the different tribes clashed with each other, and were sometimes played against each other by Taiwan's revolving door of colonizers. They have since become more united. Their solidarity is driven by their shared connection to the land and their collective history of oppression during four centuries of serial colonization.

The ROC government's apology to Indigenous Taiwanese was one of the first things Tsai Ing-wen did after becoming president. I had the chance to speak with her in October 2024, shortly after she had finished her two four-year terms. I asked Tsai how she felt during the apology ceremony.

'I was a bit nervous, because that was the first time a president of Taiwan had apologized to the Indigenous people,' Tsai told me in her British-accented English. 'Their treatment as second-class citizens is just such a sad story.'

Tsai said that her partial Indigenous heritage makes her feelings towards Indigenous Taiwanese more than just general sympathy. 'I was nervous, but I was also excited in the sense that I was able to do something for my brothers and sisters.'

While there is a slow but growing acceptance of Indigenous people in Taiwanese society, they have also taken on a new political importance in the country's democratic era. The story of Taiwan begins with its Indigenous people, who, as the island's original stewards, are also living refutations of the CCP's irredentist claims on the island country.

The rhetorical power of Taiwan's first peoples was on full display in January 2019, when leaders from two dozen tribes (not all of which are recognized by Taiwan's government) issued a letter to Xi Jinping, who, days earlier, gave a speech in Beijing to the people of Taiwan. Speaking at the Great Hall of the People on the second day of the new year, Xi delivered a hyper-nationalist speech in which he referred to Taiwanese as 'compatriots', 'friends' and 'family', while also refusing to rule out using military force to annex their homeland. Stating that 'Chinese don't fight Chinese' – a risible statement to anyone familiar with Chinese history, including the refugees who arrived in Taiwan with the KMT in the 1940s – Xi blamed Beijing's inability to solve the 'Taiwan question' on a handful of Taiwanese 'separatists' aided by 'external forces'.[3]

A fortnight later, the descendants of Taiwan's first peoples published their open letter to China's authoritarian leader.[4]

'Mr Xi Jinping, you do not know us, so you do not know Taiwan,' the letter began. 'We are the Indigenous peoples of Taiwan, and we've lived in Taiwan, our motherland, for more than 6,000 years. We are not so-called "ethnic minorities" within the "Chinese nation".'

The letter continued:

> The stories our ancestors tell of the mountains of Jade, Alishan, Dabajian, Kavulungan, Beinan and Dulan, [as well as the] forests, grasslands, valleys, rivers, islands and oceans of Taiwan testify that Taiwan is – and has always been – the traditional territory of the Indigenous peoples on this land.
>
> Taiwan is the sacred land where generations of our ancestors lived and protected with their lives. It doesn't belong to China.
>
> We, the Indigenous peoples of Taiwan, have witnessed the deeds and words of those who came to this island, including the Spanish, the Dutch, the Koxinga Kingdom, the Qing Empire, the Japanese and the Republic of China.
>
> We signed treaties with the Dutch and peace agreements with the Americans. We have fought against imperialism and every foreign intruder of our land. We have suffered military suppression from colonial and authoritarian regimes.

> *Once called 'barbarians', we are now recognized as the original owners of Taiwan.*
>
> *We, the Indigenous peoples of Taiwan, have pushed this nation forward towards respect for human rights, democracy and freedom. After thousands of years, we are still here.*
>
> *We have never given up our rightful claim to the sovereignty of Taiwan [. . .] Nevertheless, Taiwan is also a nation that we are striving to build together with other peoples who recognize the distinct identity of this land. Taiwan is a nation accommodating diverse peoples trying to understand each other's painful pasts, as well as a nation in which we can tell our own stories in our own languages, loudly.*[5]

In a world that has long been told to only see the historical, cultural and linguistic connections between China and Taiwan – which are very real – it is worth considering that, if the Indigenous Taiwanese pursued the same expansionist thinking of Beijing, they could lay claim to a vast territory. The Austronesian Expansion that took place from around 3000 to 1500 BCE was a major moment in human migration. It saw master boatbuilders and navigators from Taiwan fan out across the Indian and Pacific oceans on outrigger canoes, taking with them their languages and cultures – and articles made of nephrite, also known as Taiwanese jade, which have been found around Southeast Asia and the South China Sea.[6] A hypothetical Austronesian empire would stretch from Madagascar to Easter Island and would include Malaysia, Singapore, Indonesia, the Philippines, New Zealand, Hawaii and many Pacific Island Nations. This empire would be home to nearly half a billion people and cover more area than any other country on the planet.

The Indigenous Taiwanese, however, merely seek a fairer position in their homeland – a centuries-long goal that would, of course, suffer a huge setback if Taiwan were to be annexed by China once more.

Hungry Ghosts

Every year in Taiwan, near the peak of the stultifying summer heat of the seventh lunar month, the gates of hell open, releasing hungry ghosts – the spirits of the wandering dead who float among the living in search of solace. This is Ghost Month.

Makeshift altars on foldable tables pop up in Taiwan's cities, rural townships and villages, in front of neighbourhood shrines, scooter repair shops, convenience stores and mom-and-pop restaurants. Laid out with offerings such as cooked meats, rice wine or skinny sticks of incense in ceramic holders, these altars are meant to placate the spirits of those who died in an undignified manner – by suicide, drowning, burning, traffic accidents, war or dire poverty. Ghost money – yellow joss paper coated in silver and gold leaf and red paint – is burned in round metal canisters. Men and women face the altar to *baibai* – show respect by clasping incense sticks in front of them or pressing their palms together and bowing their heads. Children observe, making their first connections to the invisible world that seems to influence the mortal realm, for better or for worse.

In Taiwan, where politeness is a paramount virtue, there is etiquette for ghosts as well. For starters, it's best to not call them ghosts. One of the many rules of Ghost Month is to refer to these restless wanderers as 'good brothers' or 'good sisters'. Other rules include not borrowing umbrellas, speaking out your full name, looking back after eating, getting married, whistling, standing under trees or swimming.[7] All of these measures are taken to prevent entanglement with these dangerous lost souls. They are not necessarily malevolent spirits, but can complicate your life with their existence.

Even foreign ghosts from the past are given consideration. In the northern port city of Keelung, the Taiwanese set out altars with baguettes and red wine at a humble harbour cemetery to offer comfort to the souls of a group of nineteenth-century French sailors buried there. The deceased met their fate in the brief war between France and the Qing in the 1880s over northern Vietnam. The war had spilled into the Pescadores – the small archipelago in the middle of the Taiwan Strait now known as Penghu – and over to Keelung.

Taiwan's successive Japanese and Chinese rulers did not like Ghost Month, which focused on those left uncared for by society. Perhaps this was self-consciously perceived as a critique of their ability to administer Taiwan, suggest the writers Albert Wu and Michelle Kuo.[8]

'Their default descriptor is "hungry," which tells us quite a bit. Deprived of basic sustenance, ghosts need human attention. They want to be fed and have nobody to feed them,' Wu and Kuo write, adding that the ghosts aren't bandits or beggars 'because they're asocial and without attachment. They're simply isolated individuals, lost among the living.'

If our world is indeed a global village, Taiwan's existence beyond the realm of the community of nations has certain parallels with the plight of the hungry ghost. Taiwan's president cannot visit most national capitals nor address the United Nations – let alone enter its buildings. While hungry ghosts are dangerous, Taiwan itself is Asia's freest democracy and an indispensable trading partner for much of the world. Taiwan has a near-monopoly on the world's most cutting-edge semiconductors, and is among the world's top twenty-five economies.

Yet even the countries most supportive of Taiwan, such as the US and Japan, have long sought to keep an official distance from their friend, for fear of entanglement in problems with China. It is not the CCP that has erased Taiwan, however, it is us: the liberal democracies, the developed economies, the supposed friends of Taiwan. For decades, the entire world – including Taiwan – was entranced by the potential opportunities of China's rapid economic growth. But an increasingly confrontational and authoritarian Beijing, and a Chinese economic slowdown, have caused many to wake up from the magic spell of 'win-win cooperation'.

We have ghosted Taiwan, and in doing so, have lost a little of our own humanity – the connection that we should feel with all our fellow inhabitants of this planet. But we have also kept a neighbour and valuable partner outside of our community. Our continued refusal to let it in from the cold may one day bring a much greater misfortune than mere angry words from Beijing.

CHAPTER 2

Ilha Formosa and its Colonizers

It was Monday night in January 2024 at the Formosa Yacht Resort in Anping, the old port district of Tainan, and the banquet room was abuzz. A heady mix of more than 500 government officials, diplomats and visiting dignitaries had descended upon the resort for the kind of event that would usually be held in the capital.

It was a special night for Tainan. Just that morning, its popular former mayor, Lai Ching-te, had been sworn in as the sixteenth president of the ROC. Breaking with tradition, Lai held his inaugural banquet in Tainan, rather than Taipei, which has been the ROC's administrative seat since 1945. The decision to host the banquet in Tainan, Taiwan's oldest and most 'Taiwanese' city, over the conventionally more 'Chinese' Taipei was deeply symbolic. This would come to signify Lai's administration's commitment to prioritize and uplift the original Taiwanese identities that existed long before the imposition of a Chinese identity. The other side of this coin has been Lai's vocal recognition of the ROC as the identity of Taiwan's government, which helps deflect accusations of seeking 'independence'.

Speaking from an orchid-covered stage in a grey suit with a purple tie, Lai expressed his appreciation to Taiwan's friends from abroad. Among them were the ambassadors of the handful of small countries that recognize the ROC government and de facto ambassadors from countries such as the US, UK, Japan and others that lack official relations with Taiwan.

'You have brought more than your blessings here, you have also given the people of Taiwan your staunch friendship, and have given strength to Taiwanese society,' Lai said. 'Thank you all so much.'

The diplomats had all taken the clean, quiet and efficient high-speed rail from Taipei to Tainan, a one-hour and forty-five-minute trip roughly the distance of New York to Washington or London to Paris. The bullet train to Tainan, the colonial crucible where the modern Taiwanese identity was forged, takes passengers past most of Taiwan's major population centres. Whizzing down the coast at 300 kilometres per hour (185 mph), passengers snooze, engage with their phones or tuck into bento they bought at the station minutes earlier. In general, everyone is patient and quiet as they board and alight. Passengers include commuters – mostly from Taiwan's west coast – with occasional foreign travellers or residents. Theft is not a concern, with passengers unafraid to leave their belongings unattended when visiting the train bathroom. After departing each station, a progression of soothing electronic notes resonates throughout the cars, followed by announcements in Mandarin, Taiwanese, Hakka and English. Out of consideration to other passengers, the four voices request that you keep your conversational volume down and turn your phone's ringer off.

Heading down the coast from Taipei, the geography is straightforward. It's nothing but towering mountains and river valleys rising to the left, and the flat sandy and muddy transition to the choppy wind turbine-filled waters of the Taiwan Strait on the right. The spaces between cities are a mix of light industry, small-scale agriculture, red temples with orange-tiled roofs, and well-maintained roads connecting dilapidated villages. Dense bamboo thickets and lanky betel palms line narrow canals that criss-cross these inter-urban spaces, as giant banyans give welcome shade next to temples and schools.

Leaving Taipei in any direction means entering Taiwan's most populous city, New Taipei City. Fully encircling the capital, New Taipei is home to four million people. The right half of the doughnut-shaped metropolis is mountainous, sparsely populated and a popular weekend destination for hikers, cyclists and river tracers. The comparatively flat and highly urbanized left half is home to the city's administrative seat and high-speed rail station in the bustling district of Banqiao. Even just getting this far from Taipei, the dominance of Mandarin as the language on the street begins to yield noticeably to Taiwanese, the trend continuing as one heads further southwards.

In the tea-growing hills of neighbouring Taoyuan, you'll find the old town of Daxi, an old Qing-era trading post that once connected Taipei with Taiwan's south, losing prominence after the Japanese built a railway through Taoyuan. Down below on the coastal plain, Taoyuan International Airport brings in returning Taiwanese travellers, foreign tourists and international businesspeople – many of whom work in the semiconductor industry.

Further south is Hsinchu, which is quietly one of the most important cities on the planet. It is the birthplace and headquarters of the Taiwan Semiconductor Manufacturing Corporation (TSMC), one of the world's most valuable companies and producer of more than 90 per cent of the world's most advanced microchips. The alchemy-like miracles produced daily by TSMC's factories with highly specialized materials and precision tools that test the limits of physics require a massive network of suppliers. Many of those suppliers are in Hsinchu as well, centred around the city's legendary science park, which is home to more than 400 tech companies. Taiwan's importance to our chip-powered world is more than a source of pride for an isolated Taiwan. It has undoubtedly strengthened the hand of the Taiwanese who seek to broaden international support for their continued sovereignty and greater global inclusion.

Like Hsinchu, Miaoli County is home to sizeable Hakka and Indigenous populations. The county seat, Miaoli City, is much quieter, giving the feeling that you are escaping the busy north of Taiwan. Also like Hsinchu, Miaoli is home to some of the steepest mountains in Taiwan, leaving much of its more rugged, sparsely populated inland areas largely cut off from the urban coast. The township of Dahu in the Miaoli highlands produces two crops of strawberries yearly that are anticipated island wide. Some bus stops there are even shaped like giant strawberries. Miaoli's geography, agrarian society and strong support for the KMT gives it the feeling of distance from the rest of the west coast. Its tongue-in-cheek nickname, the Kingdom of Miaoli, jokingly implies it is independent from Taiwan itself (it is not).

Halfway down the coast is the sprawling, wealthy city of Taichung. It's now the country's third most populous city after being a manufacturing powerhouse during the 1980s and 1990s, when Taiwan was one of the four Asian Tigers, alongside Hong Kong, Singapore and South Korea. In the late 1990s, many of the manufacturing businesses that once operated in Taichung moved over to China where land and labour were cheaper. Taichung boomers, like those in other major Taiwanese cities, benefitted from stock and property markets and accrued substantial wealth. With more money has come an increased focus on health and leisure – the Taiwanese have increasingly taken to bicycles to explore the stunning coastal roads and mountain vistas of their country. More often than not, these cyclists are riding mid-range to high-end bikes made by Giant. Taichung-based Giant is the world's largest producer of bicycles and one of the few Taiwanese brands that has made a global name for itself. In addition to bikes, Taichung is also known for boba, the cold and sweet milk tea with chewy tapioca balls that was invented there, and is enjoyed around the world.

As you travel down the country, the train quickly passes through the smaller counties of Changhua and Yunlin, both primarily agricultural areas that saw some of the earliest Chinese settlers during the late seventeenth century. The name of Changhua's old port town of Lukang translates to 'deer harbour', as it was a major trading point for the hides of the deer that populated the once heavily forested plains of western Taiwan. Rice and other agricultural goods brought wealth to Lukang, especially to the eminent Koo clan who would move to Taipei in the late 1800s and cultivate close ties with both Taiwan's Japanese and, later, Chinese rulers. The Koo family is now one of a handful of heavily intermarried families at the top of Taiwan's economy, many of whom trace their roots back to these small rural counties.

Shortly after passing into Chiayi County, southbound travellers cross the Tropic of Cancer, leaving Taiwan's subtropical north for its tropical south, where coconut palms and tropical fruits like pineapples, mangoes, passionfruit and lychee become more prominent. Chiayi was a major source of timber for the Japanese Empire, which built a train line up into the eighteen mountains collectively referred to as the Alishan Range. Winding more than 70 kilometres into the mountains, its completion in 1914 was an impressive achievement at the time, given the steep topography. At 2,200 metres (7,190 feet), the terminus is now one of the more popular travel destinations in Taiwan's south, and is home to Taiwan's highest 7-Eleven. Down below on the Chiayi coast is the Southern Branch of the National Palace Museum. Unlike its counterpart on Taipei's northern outskirts, this branch of the museum highlights the fluidity of identity over time in Asia, a not-so-subtle political rebuttal to the northern branch's historical promotion of Chinese identity. One of the most consequential culinary inventions of the twentieth century was developed by Chiayi native Go

Pek-hok, a Hoklo Taiwanese better known by his Japanese name, Ando Momofuku. His invention: instant noodles.

The bullet train's final destination, Tainan, is where Taiwan's pre-colonial history ends and its four centuries of anti-colonial struggle begins. There are Dutch- and Qing-era forts, Taiwan's first Confucian temple, former British trading houses and a Japanese-era department store that is a major draw for shoppers. The scent of incense lilts throughout the streets and narrow, winding alleys of Tainan, which has the highest concentration of temples in the country, fusing Buddhism, Taoism and Chinese folk religion. Over the centuries, Tainan has become a walkable museum of the external forces that have shaped modern Taiwan, which themselves have been absorbed into the evolving concept of Taiwaneseness.

It is a common complaint in the country's south that development has disproportionately favoured the north, especially in the period before democratization. Many Taiwanese deride Taipei as *Tianlongguo* – the 'Celestial Dragon Kingdom', characterized by power, arrogance and aloofness, and Lai's banquet in Tainan was a message to the rest of Taiwan beyond the capital that it would not be forgotten.

'Aside from the fact that it is the four hundredth year since Tainan's founding, we chose to hold Taiwan's state banquet in Tainan because we want to show that the new administration is capable of achieving balanced development in building the Taiwan of the future,' Lai said. Taiwan, he said, should be a place where 'everywhere is a good place to settle down, and anyone can work in peace and contentment'.[1]

As with previous inaugural banquets, everything had political significance. The banquet's menu reflected Taiwan's multicultural society, with dishes in the eight-course affair representing its five major ethnic groups: the Indigenous Taiwanese, Hoklo, Hakka and

Southeast Asian populations – as well as the Chinese who escaped Mao's Communist revolution in the second half of the 1940s.

The hors d'oeuvres platter, for example, told the story of modern Taiwan in just five bites. The ingredients of each mouthful highlighted the culture of one of the five different groups that have shaped the country's diverse society.

A wrap of fragrant millet and pork represented Taiwan's Indigenous peoples, who were hunting, farming and making millet wine for millennia before the other ethnic groups showed up. While only 2 per cent of Taiwan's population is officially Indigenous, many Taiwanese from all walks of life have some degree of Indigenous blood. Indigenous flavours, languages and art permeate Taiwanese culture, and the people to whom this cultural bounty belongs also happen to be the living embodiment of why Taiwan is not historically 'Chinese'.

A spring roll stuffed with soft, dense taro and sweet potato flavoured with the tart calamansi fruit was a deep-fried nod to the Hoklo – Taiwanese who descended from migrants from the southern part of Fujian Province in China. The Hoklo made up most of the first wave of migration from China to Taiwan beginning in the 1620s, when the Dutch rapidly colonized much of Taiwan's coastal plains, importing agricultural labourers from across the strait as China's Ming Dynasty was collapsing. Tainan was where the Dutch first landed, exactly 400 years prior to Lai's inauguration. For thirty-eight years, Tainan would serve as the headquarters of the Dutch – and later their successors, the house of Koxinga and the Great Qing, under which a greater amount of Hoklo migration would occur. With the Hoklo would come the Hokkien language, which, in addition to its homeland in Fujian Province in China, is spoken by Southeast Asia's Fujianese diaspora. The variant of the Hokkien language spoken in Taiwan contains loan words from Indigenous languages, Dutch and Japanese and is commonly known as Taiwanese. It is almost completely unintelligible to someone who speaks only Mandarin Chinese, a much younger language with fewer tones.

The Hakka appetizer was a jiggly glutinous rice ball coloured deep olive green by Chinese mugwort. Following the arrival of the Qing in Taiwan in 1683, many Hakka from the provinces of Fujian and Guangdong also made the dangerous journey across the strait. In Taiwan, the agrarian Hakka traditionally occupied a social space between the Indigenous and Hoklo and politically they have tended to punch above their weight. Taiwanese Hakka are less than 20 per cent of the population, but three of Taiwan's five elected presidents (Lee Teng-hui, Chen Shui-bian and Tsai Ing-wen) have been of Hakka descent.

Drunken chicken represented the mainlanders, or *waishengren*, as those who arrived from China with the KMT (in the late 1940s) and their descendants are known in Taiwan. In this classic cold dish from Shanghai – Chiang Kai-shek's power base when he ruled China – chicken is steamed and brined in Shaoxing wine, a spirit originating near the Generalissimo's birthplace in Zhejiang Province.

A plump, shelled shrimp glazed in a neon orange sweet chilli sauce symbolized the Southeast Asians who constitute the country's newest migrant group. Taiwan has been slow to recognize the significance of Southeast Asians to its society, but that is changing. The majority of this new migrant wave comes from four countries: Indonesia, the Philippines, Vietnam and Thailand. The language of food is helping citizens of Taiwan's Southeast Asian neighbours integrate into Taiwanese society with a proliferation of eateries popping up throughout Taiwan in recent years, even in small towns.

'They're an important part of Taiwan,' said Jewel Tsai, one of the banquet organizers, of the Southeast Asian community. 'I believe that in the next five to ten years, their food will be integrated into Taiwanese cuisine at large.'[2]

The penultimate course was a simple bowl of shrimp rice and a glass of boba: sweet milky tea containing chewy black tapioca pearls that are hoovered up through extra-wide straws. On the surface, this may appear to be a simple show of appreciation for Taiwanese soft power. After all, boba was invented in Taichung and is wildly popular across the world. But the combo was also a playful reference to the elephant that is always in the room in Taiwan – the unrelenting assertion of the CCP that it is the legitimate ruler of Taiwan. As a candidate on the campaign trail, Lai had criticized China for its increasingly aggressive behaviour towards Taiwan, which has been the main driver of rising regional tensions. Lai told reporters he hoped to meet with Xi – without preconditions – saying he'd treat China's modern emperor to a simple meal of boba and shrimp rice.[3] Such a meeting is, of course, extremely unlikely.

The belligerent and stalker-like nature of China's approach to Taiwan was made awkwardly clear at a 2020 party at the Grand Pacific Hotel in Suva, Fiji. It started with a reception thrown by the Taiwanese foreign ministry for Fijian government officials, academics and NGO workers to celebrate the National Day of the ROC. Things were pleasant – until Chinese diplomats arrived uninvited. The Chinese party-crashers began to take photos of guests without permission and, when blocked by Taiwanese officials, became violent. Taiwan's foreign ministry told reporters that one of their officials was beaten so badly they had to be hospitalized. Beijing shrugged, saying their officials were simply doing their jobs, while complaining that the flag of Taiwan's ROC government had been depicted on a cake at the event. China's response was tantamount to affirming violence as a reasonable response to anyone acknowledging the existence of Taiwanese sovereignty – even on a cake.[4] For the CCP, no recognition of Taiwan's real and existing sovereignty can be brooked, no matter how small.

Beijing took things a step further in August 2024, when it passed a law that made advocating Taiwanese sovereignty punishable by death in the most serious cases.[5] The move has increased pressure on the remaining Taiwanese living and doing business in China to leave or face growing pressure to aid the CCP's efforts to subsume their homeland.

Xi and the CCP have been consistently hostile to Lai and his predecessor, Tsai. Xi has rejected multiple calls by both Lai and Tsai for the two sides to talk 'as equals and with dignity'. While pretending Taiwan's presidents don't exist, however, Xi and other top Communist officials have been happy to meet with members of the KMT, their former mortal enemies. The foundation for the current warm ties between the two parties is that both claim that Taiwan is part of China. They are also both pro-unification, Han Chinese ethnonationalist parties who see Taiwaneseness as a subset of Chineseness. Lai and others on the 'green' side of the Taiwan political spectrum (the KMT being on the 'blue' side) see Taiwan as an already independent country that has an undeniable right to exist – and could even get along with China.

The Claim

So, what is justification for the PRC claiming to be the sole legitimate ruler of Taiwan, an island nation it has never controlled?

In a January 2025 opinion piece in the *Australian Financial Review* entitled 'Seven truths on why Taiwan always will be China's', China's ambassador to Australia, Xiao Qian, provides a glimpse into Beijing's arguments supporting its claim over Taiwan. However, the article, like many of China's assertions about Taiwan, ranges from inaccuracy to delusion.

Xiao begins by claiming that Taiwan's Indigenous people are 'descendants of the ancient Bai-Yue lineage who migrated from the Chinese mainland 30,000 years ago', suggesting prehistoric human migration somehow has a bearing on modern national borders. He goes on to say that Chinese historical references to Taiwan 'date back to China's Three Kingdoms Period in 1335, making them the earliest records in the world, the imperial central governments of China all set up administrative bodies to exercise jurisdiction over Penghu and Taiwan'. Putting aside the fact that China's Three Kingdoms Period was not in the fourteenth century, but between 220 CE and 280 CE, the earliest widely accepted first-hand written account of Taiwan is by the Ming Dynasty scholar Chen Di, who visited briefly in 1603. It should be noted that there are earlier accounts of places that might be Taiwan, but could possibly be referring to other islands in the region, such as the Ryukyu Islands of Japan or Luzon in the north of the Philippines. Regardless, it should not be considered controversial to note that isolated travel accounts from centuries ago do not form the basis for contemporary territorial claims.

Xiao's claim that all Chinese dynasties established administrative control over Penghu and Taiwan is patently false. Many dynasties governed only parts of what Beijing controls today, with some closer to Central Asia than to the Taiwan Strait. Moreover, for most of its history, the territory now administered by the PRC was ruled simultaneously by different kingdoms whose borders were highly fluid. The Manchu empire of the Great Qing was the first from the Asian mainland to establish administrative offices and a permanent military presence on Taiwanese soil, and even then, only in the late seventeenth century. Manchu control of Taiwan was limited to its western coastal plain, with the Qing eventually ceding sovereignty to Japan in 1895. Notably, a Japanese government was the first to control all of Taiwan fully, which Xiao's distorted history omits. Xiao falsely maintains that 'China's sovereignty and territorial integrity have never been undermined, and Taiwan has always been part of Chinese territory'.

Pivoting to the modern era, Xiao contends that 'the international community recognizes Taiwan as an inalienable part of China'. However, the UN resolution that recognized the PRC in 1971 does not even mention Taiwan, contradicting his assertion.

The absence of a legal foundation for Beijing's claim has led to the CCP to stress the narrative of Taiwan as 'an inalienable part of China since ancient times'. For the vast majority of its human history, Taiwan was inhabited only by Indigenous peoples with distinct languages and cultures, and their own territorial battles.

Taiwan Enters the Historical Record

The first historical record of Indigenous Taiwanese people was written during the late Ming Dynasty by the Chinese military adviser and scholar Chen Di, who accompanied a punitive expedition to the island in early 1603. The expedition was led by the Chinese general Shen Yourong. Shen sought to disrupt the operations of the Wokou, ethnically mixed pirates who had established themselves on Taiwan's west coast and would conduct raids on China's southeastern coast. Shen achieved his mission, breaking up the pirate networks on Taiwan. At the time, Chinese referred to both Taiwan and its Indigenous inhabitants as Dongfan (東番, 'Eastern Barbarians'). Chen wrote an account of his experiences on the other side of the strait entitled *Record of Dongfan*, totalling under 1,500 Chinese characters. He notes that villages tend to have around a thousand inhabitants and often go to war with each other, taking heads in the course of bloody battles. The following day, however, the villages would treat the previous day's violence as water under the bridge and return to peaceful relations. Chen describes the flora and fauna of the island, as well as the social customs of the people, who are believed to be the Siraya who inhabited what is now Tainan. While Chinese fishermen and pirates had certainly made stops in Taiwan prior to Chen's visit, his record of his encounter with 'Dongfan' highlights how foreign the island was to even educated Chinese in his day.

Despite Taiwan's geographic proximity to both China and Japan, it took time for either side's rulers to show any desire to incorporate it into their respective realms. Even the Europeans were slow to take interest. Portugal and Spain would be the first European colonial powers to arrive in Asia. On India's south-west coast, the Portuguese established a protectorate in Cochin (now Kochi) in 1503, later conquering Goa in 1510. The capture of Malacca would follow in 1511. In 1543, a China-bound vessel that was blown off course made the Portuguese the first Europeans to reach Japan, where they would later set up a trading post at Nagasaki. There is no evidence that Portuguese sailors stopped in Taiwan at that point, but as they sailed past the verdant, mountainous island in 1544, they added it to a map under the name *Ilha Formosa* – beautiful island. 'Formosa' would be how most of the Western world would know Taiwan until the Cold War. Portuguese traders would continue to focus on opportunities with Taiwan's larger neighbours, expanding their ties with Japan and eventually leasing the southern Chinese port of Macau from the Ming Dynasty in 1557. The crew and passengers of a Portuguese ship became the first known Europeans to set foot on Taiwan's shores in 1582 after their vessel ran aground on a sand bar. The determined and resourceful group was stranded there for forty-five days, during which they suffered from malaria and fended off attacks from Indigenous Taiwanese, all while building a new boat, which they then sailed back to Macau.[6] Meanwhile, the Spanish were busy with their colonial enterprise in the Philippines, which was centred around Manila from 1571 onward. The galleons connecting Acapulco and Manila would reshape economies, cuisines and cultures on both sides of the Pacific and beyond.

Other European countries wanted access to Asia, among them the Dutch, who were still fighting for independence from Spain. At this time, maps with detailed information on coastal features, currents, sandbars and other hazards to ensure safe passage through the seas between Europe and East Asia were among the most precious of intellectual property. Both the Portuguese and the Spanish closely guarded these invaluable trade secrets.

Enter Jan Huygen van Linschoten. As an adventure-seeking twenty-year-old, the young Dutchman was swept up in what would be known in Europe as the Age of Discovery – but for Asians would be an era of colonial invasion. In 1583, he took up a clerking position in Goa under the Portuguese bishop there. During his nine years in Goa, he took detailed notes of everything he observed – while also meticulously copying Portuguese maps, some of which had been classified for more than a century. Van Linschoten returned to the Netherlands in 1592, where, four years later, he published his book *Itinerario*. In addition to chronicling his personal experiences in India, the book also included the priceless maps he had purloined from the Portuguese. The book was a huge success and the Iberian stranglehold on European navigational knowledge of what we now call the Indo-Pacific was broken.[7] Dutch vessels promptly raced outwards in search of Asian riches in such numbers that, in 1602, local authorities decided only to permit fleets approved by the newly formed Dutch East India Company (*Vereenigde Oostindische Compagnie*, or VOC) to venture forth from Dutch ports.

The Company, recognized as the world's first joint-stock company, served as a proxy for the Dutch government and its interests. By 1619, the Company had established a presence on the northern coast of Java at Batavia – present-day Jakarta – that served as its headquarters for a constellation of smaller trading posts it had built up in the region. This was not enough for the profit-focused company, which craved access to Asia's biggest and most prized market: China. In 1622, the VOC attempted to dislodge the Portuguese from Macau, an ambitious move that ultimately failed. The persistent Dutch turned to the Pescadores, now known as Penghu. They established a fort overlooking a deep bay adjacent to a large Chinese temple, employing Chinese to do backbreaking construction work in the tropical heat. From this base, they hoped to engage in trade with the Chinese coast. In 1624, however, the Tianqi Emperor sent thousands of Ming troops to the rocky islands. They arrived bearing an ultimatum for the *Hongmao*, or 'Redhairs' as they referred to the Dutch: relocate beyond Chinese territory, or else. The Dutch complied. They dismantled their stronghold on the Pescadores and built a new fortress just 50 kilometres away on a sandy stretch of Taiwan's south-western coast.

As historian Tonio Andrade notes in his book, *Lost Colony*:

> Since the Chinese government didn't claim Taiwan, the Dutch dismantled their fortress [on the Pescadores], sailed to Taiwan, and, on a long narrow tongue of land just off the southern shore, constructed a new fortress, which they called Zeelandia Castle.[8]

When the Dutch arrived, there was certainly no evidence of Chinese administration or authority. Of course, the area around Zeelandia Castle was not uninhabited. There were Indigenous villages further from the coast, whose men took heads – from other tribes or unlucky outsiders – as trophies. There were also Fujianese pirates and Japanese traders with which to contend. Despite being new arrivals in unfamiliar territory, the Dutch did have one major advantage that was key to their success: state-of-the-art muskets and cannons.

The Dutch relied on more than brute force – they could be pragmatic as well. Rather than expel or enslave the Fujianese pirates, they invited them to settle in the new colony and even employed them. Despite the corporate imprimatur, the Dutch East India Company was effectively a better-organized group of pirates that were happy to employ their buccaneer brethren to plunder, or at least disrupt, rival Portuguese and Spanish vessels that were plying nearby waters.[9] In 1626, the Spanish established *Hermosa espanola*, a colony in Taiwan's north, primarily to keep an eye on the Dutch. With their capital in Keelung, the Spanish engaged in proselytizing and sulphur mining, among other endeavours, and were eventually expelled by the much more established Dutch in 1642, only seventeen years after arriving.

Unable to interest local tribespeople in commercial farming, the Dutch looked northwards to China, where the Ming Dynasty was falling apart and cheap labour was easy to come by. The initial migration of Chinese to Taiwan began with the first Chinese to come over from the Pescadores in 1624, with demand for field workers growing as more arable land was opened up. Male labourers from China's south-east coast, from the provinces of Fujian and Guangdong, would follow in search of opportunity.

Most of the men who arrived in Taiwan from China were Hoklo from southern Fujian. The arrival of the Manchu Qing Dynasty in 1683 led to the beginning of cross-strait migration by the Hakka – a distinct group in southern China, thought to have originated in the north centuries earlier. Today, after the Hoklo, Hakka are the second-largest ethnic group in Taiwan. The Hakka tended to live on the periphery of Hoklo settlements and interacted extensively with local Indigenous tribes. Hakka communities today are primarily concentrated in mountainous interior areas of the north-western counties of Hsinchu and Miaoli, as well as Kaohsiung and Pingtung in the south. Intermarriage between migrants from China and Indigenous Taiwanese was not uncommon, with some marriage records showing that many Dutch men also married Indigenous women. As an Indigenous heritage was not socially advantageous, many people born to mixed couples hid their Indigenous roots.

As history shows, far from conquering existing Chinese settlements, the initial success of Dutch colonization in Taiwan was partly due to the *lack* of interest from China (and Japan) in the island. But this would change after the arrival of Koxinga.

CHAPTER 3

In the Shadows of Empire

Born in 1624, the same year as the founding of the Dutch colony at Tainan, Koxinga, better known in Mandarin as Zheng Chenggong, was a fearsome general of the Southern Ming. The son of Zheng Zhilong, a Fujianese pirate of northern Chinese extraction, and Tagawa Matsu, Zheng Zhilong's Japanese wife, Koxinga had greater ambitions than to be a pirate. He wanted a kingdom of his own.

Initially indifferent towards Taiwan, Koxinga sought to establish his own kingdom in China. As a young general, he led Ming troops to fight the invading Manchus, earning him a reputation for his loyalty to the Han Ming. As the fortunes of the Ming Dynasty waned, Koxinga pragmatically shifted his focus to consolidating power and building a formidable personal army. He did this on the small island of Kinmen, 5 kilometres off China's Fujian coast.

When the Manchu Qing Dynasty sent ships to attack Koxinga and his troops at Kinmen, they were routed, partially due to a major storm that took place during the battle. Koxinga then aimed to exploit Qing vulnerability in China's south, setting his sights on the Yangtze River. He established a base at the island of Zhoushan, just south of Shanghai, where the yawning Hangzhou Bay meets the East China Sea. In his sights was the historic Chinese capital of Nanjing further upriver. Despite making significant headway and instilling fear in the Qing leadership, Koxinga's attempt to take Nanjing was foiled by his hubris, strategic errors and a reluctance to fully besiege the city. He was forced to retreat.

In the wake of this setback, Koxinga turned his gaze towards Taiwan. There were many reasons for Koxinga to look southwards: among them the island's proximity to China, economic development under Dutch supervision and large population of Hokkien speakers. But Koxinga also had a personal connection to the island. His father had worked as a translator and mediator for the Dutch, first on the Pescadores in the Taiwan Strait and later relocating to Taiwan following the warning from the Ming emperor. Koxinga himself, however, knew very little about Taiwan. That changed in 1659 when He Bin, a disgruntled former employee of the Dutch, showed up at Koxinga's new base in Xiamen. He brought a detailed map of Taiwan to Koxinga, who used it to plan his invasion of the VOC's colonial holdings.

In 1661, Koxinga attacked the Dutch settlement in Tainan, first destroying Provintia Castle further inland and then laying siege to Zeelandia Castle for nine months, during which he persuaded the Indigenous locals to join his side. In 1662, the last Dutch governor of Taiwan, the Swedish nobleman Frederick Coyett, signed an instrument of surrender that allowed for his escape after the death of more than 1,500 Dutch colonists. Coyett would be blamed by the Dutch for the loss and sentenced to execution, which was then commuted to exile. In his 1675 book, *Neglected Formosa*, Coyett would later criticize the VOC leadership for not supplying its Taiwan venture with the troops and resources needed to repel Koxinga and his forces.[1]

With his successful conquest, Koxinga named himself ruler of a Taiwan-based dynasty that he called Tungtu, or 'Eastern Capital'. But his rule didn't last long: he would die of malaria just a few months later at the age of thirty-nine.

Koxinga's son, Zheng Jing, succeeded his father. Zheng, also an accomplished general, would rename the dynasty Tungning, or

'Eastern Pacification'. Centred around Tainan, the dynasty would last only twenty-two years, compared to the thirty-eight years of Dutch rule. During its two decades, however, it did much to Sinicize the settlement, building Taiwan's first Confucian temple and academy and opening the first language schools that taught Chinese to local Indigenous peoples. As more Chinese male labourers arrived, mostly from Fujian, Tungning expanded the scope and efficiency of agriculture and industry in Taiwan. The dynasty cleared more land for rice paddies, introduced salt drying through evaporation (a technique still used in southern Taiwan today) and expanded sugar cane farming and industrial sugar refining. Just as the Dutch brought mango, lychee and wax apple seedlings to Taiwan from its colonial holdings in the region, the Tungning dynasty introduced plum and peach trees from China.[2] Today, the plum blossom is the national flower of the ROC in Taiwan and is used in the bizarre Olympic flag forced upon Taiwan's 'Team Chinese Taipei', which also incorporates the KMT emblem and the Olympic rings.

The Manchus previously thought little of the island, with the Kangxi Emperor famously saying that, 'Taiwan is no bigger than a ball of mud. We gain nothing by possessing it, and it would be no loss if we did not acquire it.'[3] It would only be with the economic success of the Tungning dynasty that the Qing would take notice and, eventually, take action against the pro-Ming colony across the strait.

In 1683, the Qing admiral Shi Lang defeated the naval forces of the Tungning. After he had spent time in Taiwan following his conquest, Shi Lang became convinced of its economic and strategic value. He made the case to the Qing court for using Taiwan to protect the south-eastern Chinese coast from outside forces – some of which had already taken interest in the island. Shi prevailed and, in 1684, the Qing officially announced it was annexing Taiwan as part of Fujian Province. As historian Emma Teng of MIT notes, the repatriation of Zheng troops and Ming royal descendants from

Taiwan to China brought the Han Chinese population in Taiwan below 100,000.[4] She writes that, following annexation and Taiwan's addition to Qing empire maps:

> *Qing imperial policy on Taiwan fluctuated between an anti-immigration policy designed to prevent Taiwan from once again becoming a rebel base, and a pro-colonization policy aimed at promoting Chinese settlement of the island and the expansion of agriculture, forestry, and other profitable economic opportunities. Whatever the government policy, population pressure and limitations on available arable land on China's southeast coast made Taiwan an attractive destination for Chinese migrants.*

In the following decades, Chinese elites in Taiwan funded the expansion of Chinese education and culture, as what had once been frontier outposts began to increasingly resemble cities.

James Lin, a historian of Taiwan at the University of Washington, told me that Qing sovereignty over Taiwan was limited, both in terms of the Han Chinese-settled territories and the Indigenous areas in which it had no presence.

'In Han areas, [the] Qing administration was able to tax and enforce laws, but as with most Qing frontier regions, had limits on its capacity,' he said.

As the Qing were viewed as foreign colonizers, they quickly found that they were not in full control. Qing cooperation with strongmen and local elites such as the Lin family of Wufeng – later one of the five biggest families under Japanese rule – was necessary to keep the enterprise together. A Qing-era gripe about the Taiwanese people summed up the challenges of subduing the island: every three years an uprising; every five years a rebellion.

'In part, why Taiwan was subject to such frequent uprisings against Qing rule was because the Qing presence was relatively weak, and the Qing relied on the military strength of local militia to help subdue threats,' Lin told me.

Qing administrators sought to control these uprisings through quarantine policies on Han settlements, which they believed pushed Han settlers to encroach further inland into Indigenous territories and inflamed tensions between Han and Indigenous Taiwanese. At other points, Lin noted, Qing administrators encouraged migration to Taiwan because of the tax revenues it could produce for the Qing state.

By the time the Manchus had shifted their view of Taiwan from irrelevant to important, the British Empire had also entered the picture. Its victory over the Qing in the First Opium War in 1842 led to the creation of treaty ports in which the British – alongside other European powers, Japan and the US – were granted extraterritoriality. In other words, the land concessions they obtained in treaty port cities became legally their territory rather than Qing territory.

The Treaty of Tientsin, which was signed in 1858 and ratified in 1860, opened up new treaty ports in China to the imperial powers, as well as two in Taiwan: Tamsui in the north and Tainan in the south. Tamsui's importance as an export hub for tea, coal, camphor, sulphur and other goods grew following its forced opening to foreign trade. There the British established a consulate at the former site of Fort Santo Domingo, which the Spanish had abandoned following their brief stay in Taiwan's north two centuries earlier. In Tainan, then known as Taiwan-fu, the British built a customs house and Western trading companies set up shop. The British also erected a consulate in the port city to the south, Takao – now known as Kaohsiung.

In addition to extractive capitalists, the opening of Taiwan to the West brought in missionaries and medical doctors, among them James Laidlaw Maxwell and George Leslie Mackay. In the late nineteenth century, the two would not only introduce modern medicine but also the Presbyterian Church to Taiwan. The church would prove to be a powerful force in Taiwanese identity politics going forward. Under Japanese and later KMT rule, the Presbyterians pushed for human rights and democracy in Taiwan, playing a major role in the ending of martial law in the 1980s and the subsequent democratization of the 1990s.

Taiwan Under Attack

The late nineteenth century also saw one of the earliest interactions between the US and Taiwan. On 19 June 1867, Rear Admiral Henry H. Bell, commander of the US Asiatic Squadron, sent a cable to the Secretary of the Navy in Washington with an update on the Formosan Expedition that he had been charged with leading. He did not have good news.

Months earlier, the Indigenous Kaolut people had killed fourteen Americans and an unspecified number of Chinese who had been aboard the merchant ship *Rover* after it shipwrecked on a coral reef near Taiwan's southern tip and they sought refuge onshore. Bell's mission was to exact revenge upon the Kaolut. Commanding two ships with 181 American officers, sailors and marines, Bell, a veteran of the US Civil War, arrived near the site of the *Rover* shipwreck on the morning of 13 June. They were welcomed with gunfire from the Kaolut. Eventually, they managed to find a safe place to anchor. The troops disembarked, taking with them enough water and provisions for four days. They would flee under extreme duress only six hours later.

Just before 10 a.m., as the punishing heat of the southern Taiwan sun intensified, Bell sent Lieutenant-Commander Alexander MacKenzie to lead a detachment to attack the Kaolut, watching from a safe distance.

'Their muskets glistened in the sun, indicating the kind of arms they carried,' Bell wrote of their matchlock firearms. 'As our men marched into the hills, the savages knowing the paths boldly decided to meet them, and gliding through the high grass and from cover to cover, displayed a stratagem and courage equal to our native Indians.'⁵

MacKenzie's group chased the Kaolut warriors through the hottest hours of the day until they were forced by the heat and steep, rugged terrain to rest. As they rested, a group of Kaolut ambushed them, one of them landing a musket ball in the chest of MacKenzie, who minutes later died under the scorching southern Taiwan sun. When the group returned to Bell with MacKenzie's body, fourteen of them were suffering from sunstroke. That evening, Bell decided his men couldn't last another day under such unwelcoming conditions.

'No sailors, indeed, no troops unaccustomed to bush life, ever displayed better spirit, but it was apparent that sailors are not adapted to that kind of warfare against a skillful enemy,' Bell wrote, noting his men achieved little, having only 'burnt a number of native huts, and chased their warriors until they could chase no longer, though at a grievous cost of life'.

In 1871, less than a decade after the *Rover* party met its fate, a shipwrecked party of Ryukyuans found themselves seeking shelter near Taiwan's southern tip. They wandered into the Paiwan village of Sinevaudjan, where locals massacred fifty-four of them after a cultural misunderstanding that remains unclear today.⁶

In 1874, three years after the slaughter, Japan, which considered the Ryukyu Islands part of its sphere of influence, invaded and levelled Sinevaudjan, massacring its inhabitants. Japan left Taiwan only after the Qing paid an indemnity for the original massacre of Ryukyuans. In return, Japan recognized Manchu control of Taiwan for the first time. Ominously for the Qing, Japan's empire expanded to Taiwan's doorstep soon afterwards, when Tokyo annexed the Ryukyu in 1879, renaming the islands Okinawa Prefecture.

The Japanese attack on Taiwan led the Qing to take Taiwan's security more seriously, but that didn't stop the attacks from other nations. In August 1884, the French launched an assault on Taiwan's north, landing more than 2,000 troops to establish a stronghold in Keelung. Two months later, an emboldened France attempted to take Tamsui but was repelled by Qing forces. The defeat was a significant blow to the French, who were unable to expand their foothold in Taiwan and ultimately vacated their tiny corner of Keelung after an eight-month occupation.

The Qing Dynasty's victory over the French may have temporarily boosted morale, but did nothing to lessen the threat posed by a rising and rapidly industrializing Japan. In the First Sino–Japanese War of 1894-95, which was mainly fought over control of Korea, the Japanese made quick work of the Qing, raising eyebrows in the West. In the Treaty of Shimonoseki, signed in April of 1895, the Qing Dynasty ceded its claim to sovereignty over Taiwan and the Pescadores to Japan in perpetuity. Japan had joined the colonial club and Taiwan would be its showcase colony – and its springboard to southern China and Southeast Asia.

Shortly after seizing Taiwan, an expansionist Tokyo was already looking further afield. To become a true Asian power, Tokyo would need to assert full control over Taiwan and further industrialize it. It would succeed and, in the process, create the foundations of modern Taiwan as we know it.

Taiwan Under Japanese Rule

When Kabayama Sukenori disembarked at the small northern port city of Keelung on 5 June 1895, the Imperial Japanese Army was waging war with the Qing-aligned Republic of Formosa that had been founded following the end of Manchu rule of western Taiwan. Spirited Taiwanese resistance to Asia's most powerful military would last another four months, ending with the capitulation of Tainan on 21 October. As a decorated figure in both the Imperial Japanese Army and Navy, Kabayama had been named Japan's first governor-general. He was tasked with consolidating Tokyo's rule over its newest and fourth-largest island. It would not prove to be an easy job.

Kabayama initially set up a provisional Office of the Governor-General in the customs office complex at Keelung. The compound had been designed by Western architects and built with local labour. He would soon move his colonial administration to Taipei (known in Japanese as Taihoku), initially staying in the Qing administrative offices, which were neither big nor imposing enough for Taiwan's new imperial rulers. The Japanese quickly went to work on building the administrative centre of Taipei that is still in use today, centred around the Office of the Governor-General, which is now the Office of the President of the ROC.

One of the first large buildings erected by the Japanese in Taipei was the Taihoku Prison. While the Governor-General's office and other administrative buildings were a hybrid of British and French architecture, the Taihoku Prison's design was distinctly American. Its hub-and-spoke design, which allowed guards stationed in the centre to see down all the corridors of the prison, was modelled after East Pennsylvania State Penitentiary. That such a prison was one of the first building projects of the new Japanese administration highlighted their awareness that not all Taiwanese would welcome imperial rule. And Japan was the most organized and advanced power to come to Taiwan yet.

Unfamiliar with Taiwan, Japan meticulously inventoried the resources, peoples and geography of its new possession. In what was probably an awkward moment, Japanese surveyors had to inform their superiors on the home islands that Mount Fuji was now the second-tallest mountain in the empire. The tallest mountain under the Rising Sun flag was now in Taiwan. Originally known by its name in the Tsou language, Patungkuonu, the Japanese would call the mountain Niitaka-san (新高山), literally 'new tall mountain'. Today it is known as Yushan (玉山) in Mandarin, which directly translated gives us its English name: Jade Mountain.

Along with exploring its new possession, Japan sought to 'pacify' the island, and become the first power to actually administer all of Taiwan. Indigenous Taiwanese rejected and fiercely resisted their new overlords. It took the Japanese two bloody decades in the rugged and unforgiving mountains of central and eastern Taiwan before they would secure full control.

Less than a decade into Japanese rule, however, changes in the mostly Han cities on the west coast were noticeable. 'SAVAGE ISLAND OF FORMOSA TRANSFORMED BY JAPANESE', a *New York Times* article dated 25 September 1904 exclaimed. The subhead elaborated: 'Wonders Worked in a Few Years With a People That Others Had Failed to Subdue – A Lesson for Other Colonizing Nations'.[7] The article praised Japan's 'taming' of the Formosans, presenting Tokyo as a 'lenient' colonizer bettering the lives of locals with new schools, train lines and modern finance:

> *Though the country had hitherto enjoyed only a few years of complete peace under Japanese rule, the appearance of the country and the spirit of its formerly savage inhabitants have already completely changed [. . .] The natives begin to understand the blessings of Japanese rule and praise it.*

The Taiwanese did not, of course, uniformly praise or even obey Japan and, just as they did under the Qing, protested from the beginning – the first major incident coming just six months after Kabayama's arrival. There would be even bigger instances of Taiwanese anger boiling over. In 1915, an armed uprising led by a man named Yu Ching-fang and his spiritual followers saw rebels, both of Indigenous and Chinese descent, storm multiple Japanese police stations. In the aftermath of what would be known as the Tapani Incident, the Japanese killed thousands, perhaps tens of thousands, of Taiwanese.[8] The last major uprising against Japanese rule took place in 1930, following a cultural misunderstanding between a Japanese policeman and members of the Seediq tribe of Musha Village. The Seediq had long been simmering with anger over the lack of respect for their culture and traditions that the Japanese had shown, and this was a tipping point. Days later, Seediq warriors raided police stations for weapons before converging on a school where they killed 134 Japanese, including many women and children. Most were beheaded. The Japanese response was swift and brutal, with 2,000 soldiers attacking Musha Village and the Japanese military later bombing Seediq hiding in the mountains with mustard gas. Months later, other tribes working under Japanese supervision attacked Musha, beheading any males they could find over the age of fifteen.

The Second World War further complicated the socio-political dynamics in Taiwan, as Japan conscripted Taiwanese men to fight against the Republic of China's army in Southeast Asia. The war also highlighted Taiwan's strategic location at the crossroads of China, Japan and the Philippines – in December 1941, shortly after the attack on Pearl Harbor, Japanese fighters caught the US military off-guard in Luzon, the largest island of the Philippines, which had been an American colony since 1898. The attack originated from Taiwan.

The American diplomat George Kerr was stationed in Taipei during Japanese rule and wrote one of the best accounts of the changeover of Taiwan from a Japanese to a Chinese holding, which is discussed in Chapter 4. He also describes how Japan used Taiwan to surprise the Americans during the war itself:

The rain of bombs on Luzon and the rattle of gunfire about Manila brought a rude awakening. Waves of Japanese bombers and fighters flew down from Formosan airfields, striking here and there along the way. Baguio was bombed at 9:30 A.M. All but two American planes were caught on the ground at Clark Field and destroyed at 12:45 P.M. On the next day the great Cavite Naval Base was put out of action. The Grand Marshal of the Philippines Armed Forces, General Douglas MacArthur, had lost his principal shield.[9]

Taiwan would remain under Japanese rule throughout the Second World War, but by late 1944, it had become clear that the Allies were heading towards an eventual victory. Japan's surrender would eventually lead to the arrival of Chiang Kai-shek and his KMT party-state, the ROC, in Taiwan. Chiang would order the removal of many Japanese buildings – most notably the Grand Shrine, a Shinto edifice that symbolized Tokyo's political control over Taiwan. The office of the Governor-General, bombed by the Americans in 1945, would be rebuilt by its next occupants, the KMT. They would rename it Chieh Shou Hall, or 'Long Live Chiang Kai-shek Hall', after Taiwan's new emperor.

CHAPTER 4

Shifting Winds, Deadly Storm

Hsieh Wen-ta was born in 1901 to a Hakka family in the central city of Taichung. It was the sixth year of Japanese rule of Taiwan, which had renamed Hsieh's hometown Taichū. That was where an impressionable sixteen-year-old Hsieh witnessed an aerial performance by the American stunt pilot Art Smith. It was 1917, and the miracle of flight was still quite new. Hsieh decided then and there that he would become a pilot.

No Taiwanese had ever become a licensed pilot, but that didn't discourage Hsieh from leaving home to attend the Itō Aviation Academy on Tokyo's southern outskirts and chase his dream. Hsieh would graduate with honours, becoming Imperial Japan's first Taiwanese aviator. He would soon demonstrate to the empire where his allegiances lied.

In the summer of 1920, Hsieh came in third place in an aerobatics competition in Tokyo, making him an instant hero back in Taiwan. That October, he put on several aerobatic shows for thrilled audiences back home. The young pilot inspired immense pride among many Taiwanese — especially students — who were seeking more significant political participation in their daily lives and the opportunity to have their voices heard in Tokyo. A local campaign tapped into this Taiwanese pride, successfully crowdfunding enough money to buy Hsieh his own plane. It was christened the *Taipei*.

One of the few surviving photos of Hsieh as a young adult shows him standing at the front of the *Taipei*, his right hand around the far edge of its propellor blade. He is in full aviator gear, his mask resting high on his forehead, pushing his black hair up at the top. In the undated black-and-white photo, Hsieh's mouth is shut tight and emotionless, his eyes are looking up and to the left of the photographer. He appears preoccupied. Perhaps he was thinking of his grandfather, Hsieh Tao-lung, who had helped lead the Republic of Formosa's resistance against the Japanese in 1895. Following the failed attempt to defend Taipei, the elder Hsieh fled across the strait to China. He would return to his Japanese-occupied homeland a year later, maintaining a much lower profile. The younger Hsieh's experience would end up echoing his grandfather's.

In early 1923, a visiting Taiwanese delegation led by the physician and activist Chiang Wei-shui presented the government in Tokyo with a petition asking the imperial government to establish a Taiwanese assembly. This would be the third such petition. Chiang was the founder of the Taiwan Cultural Association, a non-violent home rule movement seeking to expand Taiwanese rights by working within the Japanese system. In addition to fostering Taiwanese nationalism, the association sought Taiwanese participation in local government and representation in the central government in order to end the Japanese treatment of Taiwanese as second-class citizens.

Timed to coincide with the delegation and their new petition, the twentysomething Hsieh flew the *Taipei* into the heart of Tokyo. Buzzing over the imperial capital, Hsieh's passenger and unknown accomplice dropped the *Taipei*'s payload on the unsuspecting civilians below. Tens of thousands of leaflets fluttered down upon the people of Tokyo, all of them voicing Taiwanese displeasure with Japanese rule. The leaflets aimed at Tokyoites' sense of moral decency, rather than advocating independence. Messages included: 'Taiwanese Have Long Been Suffering Under Tyrannical Rule', 'The Totalitarianism of the Colonial Government is a Disgrace to the Constitutional Country of Japan!' and 'Grant Taiwan a Representative Assembly'.[1]

Hsieh and the petitioners didn't succeed in securing democratic representation for his people. But his daring flight across the metropole provided a boost to the evolving Taiwanese consciousness and growing desire for self-determination – or at least some degree of self-administration. It was also a major embarrassment for the ascendant Japanese empire, who hadn't dealt with large-scale Taiwanese insubordination since the 1915 Tapani Incident in Tainan. Since then, much of Tokyo's focus had been on bringing the island's Indigenous peoples and the vast mountainous terrain they inhabited under imperial control.

Hsieh's leaflet stunt put the aviator in hot water with the dour-faced and moustachioed Baron Den Kenjirō, the first civilian governor-general of Taiwan. The baron chided the popular pilot for associating with Taiwanese democracy advocates and made a not-so-veiled accusation that Hsieh was attempting to break up the empire,[2] but allowed Hsieh to walk out of the Governor-General's Building a free man. Hsieh's popularity with the Taiwanese people may have helped him evade serious punishment for the time being, but he wasn't taking any chances. He fled across the strait, from Japanese Formosa (as Taiwan was known) to the ROC. He would stay in China for two decades, initially as an ROC Air Force pilot and later as a businessman and anti-Japan activist.

During a half-century of Japanese colonial rule, hundreds of thousands of Japanese moved to Japanese Formosa, where they brought infrastructure and industrial development to the island. Education rates soared among Taiwan's population, including girls and young women. Health outcomes improved. Scientists developed ponlai rice, a strain of short-grain rice that could survive Taiwan's heat. It is still the dominant variety in Taiwan today, and is used in brewing Taiwan Beer. The Japanese also introduced Taiwan's national pastime of baseball, which is even featured on the NT$500 note. As Taiwanese lives improved materially, however, the Japanese empire was becoming overstretched as it fought in vain to thwart the Allied advance across the Pacific Island and Southeast Asian territories whose conquest had been facilitated by Taiwan's strategic location.

The American-led bombing of major cities across Taiwan on 14 November 1944 was a harbinger of the end of the war. Allied planes would continue to regularly menace Taiwan's skies until Japan's surrender in August 1945. In May 1945, the US diplomat George Kerr wrote, the administrative centre of Taipei was destroyed by what he described as a 'fire-carpet', while rail centres, landing strips and hangars island-wide were pockmarked with craters.

'The harbors were choked with burned and capsized ships,' he recounted, adding that the two main port cities of Keelung and Takao (Kaohsiung) were 'virtually wiped out'.[3]

In addition to bombs, the Americans took a page from Hsieh's book, dropping hundreds of thousands of anti-Japanese leaflets upon cities and towns below. The messages, written in both Japanese and Chinese, urged the Taiwanese to refrain from supporting Japan's war effort, while also offering the promise of liberation via a 'return' to Chinese rule under the ROC. The handbills also included the preamble of the United Nations Charter. Less than one year after Japan had belatedly promoted Taiwan to a full prefecture with representation in Tokyo, the Taiwanese people were now looking with cautious optimism towards a new future.[4]

The Transfer of Power

In the fifty years it experienced under Japanese rule, Taiwan underwent significant infrastructural and economic development. It went from being a poorly connected, mostly illiterate peasantry to a prosperous, effectively administered, highly educated, connected and modernized society with advanced transportation and health care. However, the island's transformation was not motivated by benevolence, but by Imperial Japan's desire to optimize returns from its first colony. Taiwan, it hoped, would serve as a showcase for a new *Pax Japonica* in Asia, an 'East Asia co-prosperity sphere' as it was called by Tokyo, with the Japanese sitting atop it all. But it was not to be. Following the end of the war, Japan relinquished control of Taiwan, which it had transformed into an economic and military prize in just five decades.

Word quickly spread in war-torn China of the shocking abundance that could be found on Taiwan, which some referred to as *baodao*, or 'treasure island' – a modern, industrialized island blessed with agricultural bounty and with a population of only five million.

As James Lin told me, 'By 1945, Taiwan was a highly productive economy, generating a healthy surplus from taxing its agricultural industries, and though it was bombed during the war, it was still seen by the KMT as a vital economic region that could be leveraged to help the mainland's economy.'

Despite this economic prosperity, the Japanese administration had imposed a system where Taiwanese were treated as subjects, with little say in their own affairs. Some concessions were made towards the end of the occupation, but these empty last-minute gestures were rendered meaningless by the arrival of the KMT in 1945.

In late 1945, the transition of power in Taiwan from Japanese to KMT control heralded a profound shift in both governance and societal development. The arrival of the ROC's first Governor of Taiwan, Chen Yi, marked a stark contrast to the era initiated by his Japanese counterpart Kabayama Sukenori half a century earlier. Unlike Kabayama, Chen Yi didn't need to take an arduous boat ride to get to Taiwan. He simply landed at

Songshan Airport in an American military plane that had flown out of Shanghai on the chilly morning of 25 October 1945. A KMT-organized parade greeted him at the airport, with children and shopkeepers on the roadside – none of whom spoke Mandarin – shouting Japanese cheers of *'Banzai!'* Later that day, Chen received signed documents of surrender from the Japanese general Ando Rikichi. Although the surrender documents did not transfer sovereignty, Chen and the ROC military quickly made clear that such details did not matter to them: they were in charge now. The receipt of surrender documents was followed by fireworks and feasting, with Chinese soldiers revelling for a solid week afterwards.[5]

The Cairo Conference

It was at the Cairo Conference in November 1943 that the US and the UK agreed that Taiwan would be 'returned' to the ROC after the end of the war. The meeting between ROC leader Generalissimo Chiang Kai-Shek, US President Franklin D. Roosevelt and UK Prime Minister Winston Churchill was the culmination of Chiang's efforts to get China a seat among the great powers. The sausage-making was done at the residence of the US Minister to Egypt, Alexander Comstock Kirk, near the Sphinx and Giza pyramids.

A widely published photo from the summit shows the three Allied leaders and Chinese First Lady Soong Mei-ling, better known abroad as Madame Chiang Kai-shek, sitting in a row of chairs on the lawn of the Mena House Hotel. In the left foreground, with stubble on the sides and rear of his balding head, a half-smiling Chiang leans slightly forward in his military uniform, his cap in his right hand on his lap. Wearing a navy suit, the American president is turned to his right, leaning in towards Chiang. Roosevelt's body language and expression suggest he was speaking with an important ally and equal – exactly how Chiang wanted to be perceived. Just days later, the group issued the Cairo Declaration, in which they stated:

> The Three Great Allies are fighting this war to restrain and punish the aggression of Japan. They covet no gain for themselves and have no thought of territorial expansion.
>
> It is their purpose that Japan shall be stripped of all the islands in the Pacific which she has seized or occupied since the beginning of the first World War in 1914, and that all the territories Japan has stolen from the Chinese, such as Manchuria, Formosa, and the Pescadores, shall be restored to the Republic of China ...[6]

This statement of intent was not legally binding under international law, nor did it involve any consultation with the people of Taiwan. Additionally, it ignores the fact that the Manchu Qing Dynasty had given Imperial Japan sovereignty over Taiwan in perpetuity in the Treaty of Shimonoseki. That treaty was signed in 1895 – nineteen years before the start of the First World War. Chinese nationalists on both sides of the Taiwan Strait cite the non-binding Cairo Declaration and the similar Potsdam Declaration of 1945 as justification for ROC or PRC sovereignty claims over Taiwan today, while dismissing the Treaty of Shimonoseki as an 'unequal treaty' similar to the treaties that followed the Opium Wars, in which the Qing gave the British control of Hong Kong and opened up the so-called treaty ports.

In 1945, following Japan's surrender to the Allies, Taipei witnessed what was supposed to be the joint Sino–American occupation of Taiwan. However, before Chen Yi even set foot on the tarmac at Songshan Airport, signs of trouble were already emerging. The month prior to Chen's welcome parade saw the less-than-victorious arrival of the first Chinese troops – who came on US naval vessels. Kerr noted that an estimated 12,000 Chinese soldiers were transported to Keelung in the north and Kaohsiung in the south, knowing that somewhere in between were 170,000 well-rested Japanese soldiers:

> *They flatly refused to go ashore. At Keelung Chinese officers begged the astonished Americans to send an advance unit overland — an American unit, of course — through the narrow valleys leading to Taipei some eighteen miles away. The Chinese officers had heard that vengeful Japanese suicide squads lurked in the hills. Only a rancorous argument forced the Chinese to accept their fate and go ashore. At Kaohsiung the Americans, eager to empty the transports, had to threaten bodily ejection of the Chinese troops before their reluctant passengers would venture "into the tiger's lair."*
>
> *It was an inauspicious beginning, made the more so because these incidents were witnessed by the Formosans. Word soon spread, and lost nothing in the telling. Formosans along the way laughed at the shambling, poorly disciplined, and very dirty Chinese troops. It was evident, they said, that the "victors" ventured into Formosa only because the United States stood between them and the dreaded Japanese.*
>
> *Much evil and many individual tragedies were to spring from these expressions of open scorn, for the mainland Chinese were losing face, dearer than life itself.*[7]

Initial Taiwanese optimism about Chinese rule had been stoked by the leaflets dropped by American planes on major cities in previous months. Now, the former subjects of the Japanese empire were seeing red flags from their Chinese 'liberators'.

When the Japanese military began to transfer its holdings to its Chinese counterparts, the Taiwanese got an initial taste of what their life under the KMT would be like. The Japanese Imperial Army handed land, facilities, equipment and food supplies to the ROC Army, and the Japanese and Chinese navies did likewise. Japan did not have a separate branch for its air force, however, so there were no air force assets — not just no planes, but also no offices, cars, dormitories or the like – to be transferred to the ROC Air Force, which Kerr described as China's 'most modern and most pampered service'.[8] The Chinese Air Force officers saw their army and navy counterparts collecting their shares of the spoils of victory, and they wanted their cut too. They responded to this perceived slight by announcing that they would seize control of hundreds of acres of land near the airport in Songshan. Much of this land was heavily populated – the Taiwanese who would be rendered homeless by the air force land grab were given just forty-eight hours to vacate. The outraged Taiwanese turned to the Chinese general in charge of the ROC and American presences, demanding that the land seizure announcement to be rescinded. Kerr said the general was surprised by the vehemence of the Taiwanese in asserting their legal rights.[9] Unlike the Chinese, people in the new province of Taiwan were educated and expected rule of law. They were also unafraid of their new rulers. The Americans put pressure on the general, who denied the air force officers their claim on the residential areas but allowed them to keep the more rural portions. The ROC Air Force officers were livid at the American interference. Thus began the fractious US–KMT joint stewardship of Taiwan. In Kerr's words:

> From the very beginning the problem of face bedeviled the Nationalist Chinese. It was apparent to all – including all Formosans – that the Nationalists were totally dependent upon the United States. They reached the island aboard American transports, and American arms and subsidies enabled them to stay. The Air Force incident set the pattern for many more to come.[10]

The Governance of General Chen Yi

After the KMT's arrival in Taiwan in 1945, Chiang Kai-shek chose General Chen Yi to serve as its first provincial governor. Any

Taiwanese who knew of Chen's decade as the governor of the south-eastern Chinese province of Fujian, just across the strait, would have been alarmed. Chiang Kai-shek and Chen Yi's fates first became intertwined in 1927. At that time, the Generalissimo, a rising military figure, had initiated an open rebellion against the internationally recognized government of the ROC in Beijing. For Chiang, the road to Beijing ran through Shanghai, so capturing Shanghai would be crucial – not just from a strategic standpoint, but also because it would allow him to leverage its vast wealth to his advantage. Chiang's seizure of Shanghai was planned and executed by Chen Yi, who struck a deal with the city's notorious triads to ensure the smooth passage of Chiang's forces into the city.

The importance of the Shanghai conquest extended far beyond military achievements. It also facilitated a significant alliance through Chiang's marriage to Soong Mei-ling at the end of 1927. This alliance combined Chiang's military power with the financial and industrial influence of the Shanghai-based Soong and Kung families. It also provided the Generalissimo with a charming English-fluent interpreter and interlocutor in his new wife. The union of the conservative Chiang with powerful industrialists also proved a winning counterweight to the left-leaning rival government that Wang Jingwei had set up in Wuhan, upriver from Shanghai on the Yangtze.

In recognition of Chen Yi's contributions to his rise to power, Chiang appointed him as governor of Fujian province in 1934. His first task was to crush an insurrection by local generals, which he did swiftly. Chen's governance soon gained notoriety for his economic exploitation and penchant for meting out beatings, torture and death to those who protested against his austerity programme, which he called 'Necessary State Socialism'. University students who rioted against Chen's administration were often subjected to torture and execution.

In late 1935, Chiang sent Chen across the strait to Taiwan, where he attended the Taiwan Exhibition – a major event with more than a million visitors. Centred in Taipei, but spread across other parts of the island, the exhibition highlighted Taiwan's rapid economic and social progress under Japanese rule. It also showed the empire's rising regional ambitions, with pavilions spotlighting areas of southern China and Southeast Asia that Tokyo would later invade from Taiwan.[11] Chen returned praising the development of the island, which under Japanese colonial rule had shot far ahead of the bleak living standards of Fujian.

A decade later, in 1945, Chen would oversee the plundering of the island he so admired. His appointment elicited protest from Taiwanese living in China, and concern from diplomats at the US State Department. Chen's initial actions in Taiwan – including a controversial conscription order which drafted young Taiwanese men to fight against Mao Zedong's Communist revolution – only fuelled the growing discontent among the Taiwanese population. An angry and educated populace argued that Japan hadn't yet renounced its sovereignty over Taiwan (which wouldn't happen until after the Treaty of San Francisco in 1952), and therefore the island was occupied enemy territory from an ROC perspective. The conscription plan was ultimately withdrawn amid accusations that it would send young, able-bodied Taiwanese men to die in a Chinese war, while weakening the Taiwanese people's position vis-à-vis the newly arrived and already unpopular ROC.

The first seventeen months of Chen Yi's governing of Taiwan were chaotic, with the economy tanking under his mismanagement. The initial contingent of 12,000 KMT soldiers was joined by reinforcements that swelled their ranks to 30,000. Poorly trained, poorly equipped and poorly paid when paid at all, the KMT soldiers had to live off the land as they had in China. Although Japan had left behind lodging which could house 200,000 soldiers, it took most of 1946 to remove KMT soldiers from the hospitals, schools and temples they'd occupied upon arrival. Wherever they set up camp was completely stripped of anything of value, down to the door knobs.¹² This was coupled with systemic efforts by KMT officers and ROC administrators to seize private property. The resulting disorder, marked by rising inflation, unemployment and crime, was a major change from the high levels of public safety and stability Taiwan experienced under Japanese rule.

The Taiwanese writer Ong Thiam-teng summarized the zeitgeist in 'liberated' Taiwan in the August 1946 issue of *Xinxin* magazine: 'We Taiwanese were extremely happy when the pillar of imperialism crumbled. We announced in unison that Taiwan had returned to the motherland with an excitement as if we could reach heaven in one step,' he wrote. The Taiwanese had deluded themselves though, he lamented: 'We are now like a ship without oars, floating in the open sea.'¹³

Ong had been employed by the Japanese government in Taiwan, which fired him for his nativist politics and would eventually imprison him. At first, Ong was optimistic about the arrival of the KMT. He served as a representative on the newly established Taiwan Representative Council, which he would quickly realize Chen Yi had created to give the illusion of some degree of Taiwanese autonomy. After joining the toothless council, he became a loud critic of KMT corruption. This stance earned him the nickname the 'Iron Councillor' among the Taiwanese, while also securing a place for himself on the KMT's hitlist.

Like Ong, many Taiwanese were initially hopeful regarding KMT rule, but in Kerr's estimation, feelings changed roughly six weeks into being unofficially absorbed into the ROC. Upon the establishment of the new provincial government with the notoriously corrupt Chen Yi at its head, corrupt officers and administrators did everything they could to seize the vast resources left by the Japanese – including food stockpiles, industrial materials and prime real estate. The KMT and their hangers-on also confiscated private property from Taiwanese people, including family homes. These 'ill-gotten assets' – as they are known in Taiwan's present domestic political discourse – added significantly to the KMT's wealth, and remain a sore subject today.[14]

In response to KMT seizures of public and private property, Taiwanese formed vigilante units to protect local interests. But as Kerr noted, they were fighting an uphill battle:

> Looting was carried forward on three levels. From September, 1945, until the year's end the military scavengers were at work at the lowest level. Anything movable – anything lying loose and unguarded for a moment – was fair prey for ragged and undisciplined soldiers. It was a first wave of petty theft, taking place in every city street and suburban village unfortunate enough to have Nationalist Army barracks or encampments nearby.[15]

Following the underpaid rank and file, Kerr writes, the ROC officer class got their cut via special depots that handled both civilian and military supplies. Lastly, Governor Chen Yi's inner circle took control of the massive agricultural and industrial stockpiles Japan had relinquished – one year later they would be heavily depleted, with Taiwan's economy descending into chaos.

The arrival of the new government also led to a significant downturn in the public health and sanitation standards that Taiwan had enjoyed under Japanese rule. The Taiwanese people, accustomed to high levels of hygiene, were appalled by the KMT soldiers' practices of public spitting and defecation. Japanese administration had long eliminated concerns about outbreaks of deadly communicable diseases, but the 'return' to China led to a breakdown of quarantine protocols and other measures that had kept islanders safe for decades. This deterioration of the health and sanitation infrastructure soon had tangible consequences. In the spring of 1946, travellers with bubonic plague arrived from China, resulting in fourteen total cases – the last previous case of plague in Taiwan had been reported in 1918.[16] The resurgence of such a deadly disease, which had been eradicated in Taiwan for nearly three decades, further eroded public confidence in the KMT.

An entry in the archives of the Institute of Taiwan History at Academia Sinica, the ROC's national academy, acknowledges the Japanese colonial government's successful epidemic prevention. In the next breath, the entry notes the return of several long-forgotten diseases starting half a year after the KMT's arrival, while avoiding blaming the new Chinese colonial government. It began with the reappearance of cholera in April 1946. The first case emerged up north in Keelung, followed by a major outbreak of the illness the next month down south in Tainan. The epidemic then spread northwards up the coast to central and northern cities. The contrast in competence between their former and new rulers when dealing with deadly disease was jarring for the Taiwanese.[17]

Economic and public health concerns aside, perhaps one of the biggest sources of disappointment for the Taiwanese was the unfulfilled promise of increased representation in Taiwan's local government. In the final decade of Japanese colonization, the Taiwanese began to experience limited democracy. They voted for district assembly members, half of whom were nominated by the Japanese government. That taste of democracy, of some degree of say over their own affairs, only stimulated the appetite for more.[18]

In May 1946, Chen Yi had agreed to the formation of a 'People's Political Council Assembly', which set out to include Taiwanese delegates in the provincial government. However, Chen's appointment of a Chinese mayor for Taipei, Huang Chao-chin, to oversee the assembly was seen as a blatant attempt to manipulate the proceedings. The initial meetings laid bare the issues that the Taiwanese sought to resolve: they did not seek to challenge the KMT but wanted meaningful representation, a return to the rule of law and a more competently run economy. The standard by which KMT rule would be judged would be the stability and prosperity of Japanese rule. However, the next meetings by the assembly, in December 1946, 'made it clear that the Government had paid not the slightest heed to warnings and recommendations set forth in May'.[19]

Despite the despondency stemming from their initial disappointment in KMT rule, many educated Taiwanese held hope that they could help rebuild a new China under the ROC government in Nanjing, then known as Nanking. Such optimism was further buoyed by American propaganda that promised freedom of the press and greater political representation. Nevertheless, the initial period of KMT governance under Chen Yi quickly became marred by its apparent incompetence and abusive practices, setting the stage for an eruption of Taiwanese anger. While there had been no shortage of confrontation between the Taiwanese and their new rulers in the first seventeen months in Taiwan, their relationship was not yet unsalvageable. In early 1947, however, that would change.

Two Two Eight

On the evening of 27 February 1947, a scuffle over a small amount of smuggled cigarettes lit the fuse of rebellion against the KMT government in Taiwan. It was one of those bitingly cold Taipei winter evenings where the humidity seems to seep into your bones. The Pegasus Teahouse in Dadaocheng (Twatutia in Taiwanese), one of the oldest parts of the city, was bustling with Taiwanese patrons. Among the cigarette- and tea-fuelled conversations, Taiwanese dissatisfaction with life under the KMT was likely a pervasive topic. Near the entrance of the teahouse, a young widow, Lin Chiang-mai, sat in the chilly night with her toddler son, selling cigarettes to passers-by. A truck driven by Chinese Monopoly Bureau officers approached the Pegasus where they began searching for contraband. They discovered that Lin was selling cigarettes smuggled from China, rather than cigarettes produced by the government monopoly that the KMT had inherited from Japan. The agents confiscated her wares while she pleaded with two of them not to deprive her of her livelihood. That was when Officer Yeh Te-ken struck Lin in the face with the butt of his gun, drawing blood from the young mother. Taiwanese onlookers who had witnessed the exchange became enraged and surrounded the two officers. Officer Fu Hsueh-tung fired a shot into the crowd, striking one person who would die later that night. In the panic and confusion immediately following the shooting, the officers fled to a nearby police station, which sheltered them from the angry crowd.

The morning of 28 February saw anger turn to action as Dadaocheng residents formed a protest march that snaked southwards to the administrative heart of the city. In a 1993 book, journalist Chou Ching recounted arriving at Taipei's north gate as the crowd approached:

> *I saw a giant banner coming my way. Written on it were eight big characters reading, 'Severely punish the killer – a life for a life.' A leather drum 2.5 chi [about three feet] across, following behind the banner, rattled eardrums [. . .] behind the drum followed a trail of innumerable people joining the petition procession. There were men and women, elderly and children, but it was primarily youth.*[20]

Having smashed the windows of a police station at the beginning of their march, the crowd, now numbering around 2,000, had not sated its appetite for revenge. Shortly after passing the north gate, it descended upon a Monopoly Bureau office, where its rage boiled over. The angry Taiwanese beat members of the Chinese staff, more than one of them fatally. They flipped cars and ransacked the office, throwing furniture and inventory into the street outside where it was piled up and set alight. In one of the only known photos of the event, a crowd is gathered around a large smouldering pile of burned debris. The three-storey Monopoly Bureau office's first-floor windows are broken and the outside of the building above the windows is blackened with soot – presumably from a fire inside. The novelist Chung Lee-ho recorded his account of the event in his diaries: 'Regardless of what it was, the crowd grabbed and carried everything they could from the Monopoly Bureau and burned it, even cars, bicycles and rickshaws.'[21]

The crowd continued to the provincial executive office building, where it sought to present a petition seeking justice for the previous night's killing, as well as other issues it had with KMT rule. As the protestors – now, in the government's eyes, rioters – amassed outside the building, KMT soldiers on the rooftop fired into the crowd, killing a handful of people and injuring dozens.

Some of the crowd ran to the provincial radio station, just a few blocks away. They overran the studio, which for decades had aired programming in Japanese, and had switched to Mandarin following the KMT's arrival less than two years earlier. Interrupting a live broadcast, a man's voice spoke to the five million people of Taiwan in the Taiwanese language, informing the island-wide audience of the KMT killings and calling for rebellion.

'Instead of starving to death,' he said, 'we should stand up and fight, root out corrupt officials and prioritize our survival.'[22]

The radio station went off the air shortly afterwards but the KMT was unable to contain the news. That night, uprisings and violence against Chinese broke out in the densely populated north.

By the next day, all of Taiwan had been consumed by the uprising. Hundreds of Chinese would die at the hands of enraged Taiwanese before local political elites, such as 'Iron Councillor' Ong Thiam-teng, reined in the anger. Order was quickly restored with Taiwanese policing the streets, as terrified Chinese stayed indoors and KMT soldiers remained in their barracks – for the time being. Suddenly, and probably surprising all involved, the Taiwanese people were running Taiwan. Among the unarmed masses who wrested control from Chen Yi's provincial government were students, workers, jobless youth, street hawkers and local business owners. Educated Taiwanese elites suddenly found themselves in the precarious position of mediating between a heavily armed KMT and a seething Taiwanese populace.

While it's safe to assume that Chen Yi wanted to retake control and unleash his wrath, what the Taiwanese people wanted in this heady and perilous moment was much less clear. Some merely wanted the KMT to reform its governance of Taiwan, while others sought a trusteeship under the UN or even the US. Others sought what few had previously considered: the foundation of a Taiwanese state.

Almost immediately, the violence that erupted on 28 February 1947 became known as the '228 Incident'. On the following day, 1 March, the Taiwanese formed a committee to investigate what had happened during the two previous days. Chen Yi's allies also joined the committee, which became known as the '228 Incident Settlement Committee'. One of the committee members, Iron Councillor Ong, drafted a list of thirty-two demands, only seven of which pertained to recent events. The rest related to general dissatisfaction with Taiwan's brief initial taste of KMT administration.

Frustration and anger among some of the Taiwanese toward the KMT had reached a boiling point, and dialogue was no longer a viable option for all. Hsieh Hsueh-hung, the founder of the Taiwanese Communist Party formed during Japanese rule, and others like her resorted to armed resistance. She led a group of guerrilla fighters, the 27 Brigade, which fought KMT troops in Taichung. But Communist membership was extremely limited – most of the people who rose up were common Taiwanese who were fed up with their homeland's backsliding under the KMT. In Yunlin County, an ophthalmologist named Tân Chhoàn-tē led a group that overtook Huwei Airport. A citizen's militia in the southern city of Chiayi also began attacking KMT positions at the local airport.

In Kaohsiung, tensions escalated when Peng Meng-chi, the commander of troops in the city, resorted to shelling civilian areas. He later ordered troops to attack protest sites outside the city hall, main train station and Kaohsiung Senior High School. His troops went on a three-day slaughter bender, prompting the Taiwanese to call him 'Butcher of Kaohsiung' to this day.

Amid this turmoil, a besieged Chen Yi sought reinforcements from Chiang Kai-shek to crush the Taiwanese uprising. Chen's men on the Settlement Committee had pushed for ten more demands to be added to Ong's original thirty-two. They included outlandish demands such as the abolishment of the KMT's Taiwan Garrison Command and the unconditional release of Japanese war criminals. The excessive nature of these new demands gave Chen Yi the perfect pretext for not only rejecting all of the demands, but for branding the Settlement Committee itself as a rebellion.

On 7 March, Chen rejected the forty-two demands. The next day, two battalions of KMT soldiers arrived at Keelung, from where they began their indiscriminate killing. This marked the beginning of the March massacre that Taiwanese simply call *ererba*, Mandarin for the numbers 228. In reality, it was a series of massacres in cities from Keelung to Kaohsiung, each leaving behind its own legacy of trauma. During the following weeks, as KMT soldiers terrorized Taiwan's cities and countryside, their goals became more apparent. As Kerr writes:

> *By March 17 the pattern of terror and revenge had emerged very clearly. First to be destroyed were all established critics of the Government. Then in their turn came Settlement Committee members and their principal aides, all youths who had taken part in the interim policing of Taipei, middle school students, middle school teachers, lawyers, economic leaders and members of influential families, and at last, anyone who in the preceding eighteen months had given offense to a mainland Chinese, causing him to "lose face." On March 16 it was reported that anyone who spoke*

English reasonably well, or who had had close foreign connections, was being seized for "examination."[23]

A *New York Times* report from Nanjing, where Chiang's grip on China was steadily slipping, quoted numerous foreigners returning from Taiwan. All of them painted a picture of carnage and terror.[24]

'Foreigners who left Formosa a few days ago say that an uneasy peace had been established almost everywhere, but executions and arrests continued,' China correspondent Tillman Durdin wrote. 'Many Formosans were said to have fled to the hills, fearing they would be killed if they returned to their homes.'

One American told Durdin that for a time, 'everyone seen on the streets was shot at, homes were broken into and occupants killed [. . .] There were instances of beheadings and mutilation of bodies, and women were raped.'

Today in Taipei, the riverside area of Machangding is where neighbourhood children learn how to ride bicycles while older people chat in the shade of banyan trees, the sweet aroma of grilled Taiwan-style sausages wafting on the breeze. In March 1947, however, it was a major execution ground. The Xindian River ran red with the blood of the Taiwanese who were executed, their bodies thrown into the river or left to rot. While it is impossible to ascertain how many Taiwanese were killed by the KMT during its March rampage, current estimates stand at around 28,000.

The radio station where the call to rise up went out on 28 February is now a museum. Inside, it highlights the targeted nature of the killings as KMT soldiers focused on eliminating locals who could mobilize Taiwanese opinion. Among them were highly educated judges, doctors, lawyers, civil servants and others who could have potentially administered a Taiwanese state. The exhibit ends with photos and names of victims, most of whom were men, highlighting the fatherless families that were left behind to grieving widows.

While the KMT had kill lists that had to be satisfied, random violence and vengeance were still very much a part of the March massacres. Dr Ira D. Hirchy, the Chief Medical Officer for the UN Relief and Rehabilitation Administration in Taiwan, would later describe horrifying scenarios that took place during this time:

> *Boys were shot down from bicycles as they rode. One man who was sitting in his home reading his evening newspaper had his money, watch and a ring removed from his person by soldiers who entered his home, and then shot him through the back. The next morning as he was being carried in a stretcher to the hospital by his family, they were shot at, even as they entered the front door of the hospital – a Canadian Mission hospital ... A working man returning home was confronted by soldiers who had him raise his hands, then searched his person. Not finding any money they ran a bayonet through his leg; then as he fell to the ground they demanded that he stand up, which he could not do. So they shot him in the head and departed. But they only shot off his ear and he was able to tell of his experiences the next day in the hospital ward. Governor Chen Yi announced over the radio that everything was at peace again, and asked all Formosans to open their shops and resume work. The next morning a half-dozen Formosans were pushing a cart of fish to market when Chinese troops opened fire on them from the roadside, killing some and wounding others.*
>
> *In the city of Pintung [Pingtung] where the inauguration of the brief people's rule was marked by the playing of the Star-Spangled Banner on phonographs, the entire group of about 45 Formosans who were carrying on various phases of local government were taken out to a nearby airfield from which, later, a series of shots were heard. A Formosan who, representing the families of these people, went to the military commander to intercede for their lives, was taken to the public square and, after his wife and children had been called to witness the event, he was beheaded*

as an example to the rest of the people not to meddle in affairs which did not concern them.[25]

By the end of March, after unleashing a wave of death and trauma island wide, Chen Yi and the KMT had retaken control of Taiwan. While state propaganda initially boasted of successfully suppressing the 228 rebellion, it would quickly switch to suppressing the memory of 228. In the thirty-six years between 1948 and 1984, major newspapers in Taiwan only mentioned 228 four times. It wasn't until the end of martial law in 1987 that the Taiwanese finally began to feel safe to discuss 228.

Taking a step back from the blow-by-blow of this formative moment in Taiwan's modern history, it is important to keep several things in mind.

The US and UN did very little to halt or condemn the campaign of terror committed by the KMT against the Taiwanese people. Prior to the atrocities of 228, there was a hope that the US or UN might intervene and perhaps assume control over Taiwan under a trusteeship. Despite this hope, the US State Department, whose anti-Communist agenda aligned with the KMT's interests, and the UN, where the ROC held the China seat on the Security Council, failed to act.

For a brief period of just over a week, the Taiwanese people had successfully overthrown the KMT and claimed sovereignty over their homeland. The political leadership in Taiwan sought greater autonomy but did not seek the outright destruction of ROC rule. If they had resolved to remove Chen Yi and the KMT before reinforcements could arrive from China, the island's five million residents might have had a chance of overrunning the ROC military forces. The understandable reluctance of Taiwanese elites to pursue full military control paved the way for a brutal crackdown that continues to cast a long shadow over Taiwan's political landscape and Taiwanese family histories.

While today's CCP has a clear carrot-and-stick strategy (with a growing emphasis on the stick) in regards to Taiwan, the PRC has never done anything remotely as bad to Taiwan as the KMT did during the 228 massacres. The atrocities committed in China during the Mao era and the carnage of the 1989 Tiananmen massacre are widely acknowledged in Taiwan today. However, they are largely seen as tragedies suffered by the Chinese people. The Taiwanese people have their own traumatic experiences with Chinese nationalism – but they're primarily connected to the KMT, not the CCP.

In essence, the events of 228 were a turning point that created the post-war Taiwanese identity. As the KMT lost its grip on China, it opted to terrorize the Taiwanese people, setting them on a path of determined resistance and movement-building that would lead, with no shortage of luck, to today's multi-party democracy in Taiwan. International media reports today refer to the KMT as the 'largest opposition party' in Taiwan – an economical and understandable response to article word limits, but one that glosses over the KMT's complex and dark relationship with Taiwan.

It was 228 that denied a generation of simple democratic freedoms. Keeping it out of public discourse was vital to maintaining the pretence of the KMT's right to rule Taiwan – everyone knew that it was American support that had allowed Chiang's KMT to commit the atrocities it did, while Washington kept its criticisms of its anti-Communist ally both muted and private.

Just as 228 was the catalyst for the oppression of the Taiwanese people from 1947 onwards, it was the eventual public discussion of 228 decades later that led to the blossoming of one of the world's freest democracies.

'The quest to redress 228 has been a very important

development,' Su Ching-hsuan, executive secretary of the Taiwan Association for Truth and Reconciliation, told me. 'It's had a major influence on Taiwan's democracy movement.'[26]

At the time I spoke to Su, a group of government-appointed scholars had just published the results of a study in which they concluded that Chiang Kai-shek was personally responsible for the horrors of 228. The smoking gun was a document dated 2 March 1947, three days after the violence outside the Pegasus Teahouse that started it all. The document was a request for at least a regiment of troops to quell the Taiwanese rebellion. Chiang personally approved the request.[27]

The Aftermath

Michael Fahey is a Taipei-based political commentator, lawyer and former journalist who, like many Americans, moved to Taiwan in the 1980s to learn Chinese, intending to eventually move on to a China that was just beginning to open up to the outside world. Fahey's arrival in Taiwan coincided with the end of martial law – a period whose presence still lingered.

In pre-internet Taiwan, learning about the darker moments of KMT rule wasn't easy. Fahey first learned of 228 while on a bike ride on Fudekeng, a small mountain on Taipei's east side that is covered with graves, many of them belonging to victims of political killings by the KMT. It was the traditional Tomb Sweeping Festival, when families visit their ancestors' graves, and, at a small temple at the foot of the climb, he encountered a group of around fifty people. They were dressed all in black and carried signs that referenced the date 2/28. Fahey and his cycling buddies paused, letting the group pass.

'A serious-looking middle-aged woman came up and asked me if I knew why they were there,' he told me. 'She said that the KMT had committed terrible crimes against the Taiwanese people and that I should know about it.'

In retrospect, he supposed they were going to visit the unmarked graves further up the hill.

'At my next Chinese class, I asked my teacher, Mrs Zhao, about 228,' he said. 'Remember there was no Wikipedia in those days. I couldn't just look it up. She calmly nodded her head and said that it was a sensitive topic but that I should stay after the group class and she would tell me.'

Mrs Zhao was from Hebei, the northern Chinese province that wraps around Beijing. She came to Taiwan after the Second World War because her husband was a scientist who had been educated in Japan – a professional advantage in Tokyo's former colony. By Fahey's accounts, Mrs Zhao was a remarkable teacher. Calm, patient, well-educated. She brought several large sheets of paper to class every day and would write complex Chinese characters for his class, upside down from her perspective, with her brush. But that day, she would give him a history lesson.

'She closed the door significantly and told me that it was still best not to discuss this issue in public,' he said. 'It seems to me that her main concern might have been the other teachers, who were 90 per cent older *waishengren* [arrivals from China].'

Mrs Zhao told him that the days after 228 were terrifying. There were riots and she had heard that people from China had been killed or roughed up. She and her husband took refuge in a Taiwanese neighbour's house.

'After a few days, order was more or less restored. But then, a few days later, the KMT army came and many people were killed,' he said. 'She told me that they committed horrible crimes.'

Mrs Zhao told him that before 228, she had had a good relationship with her Taiwanese neighbours – she had even become conversant in the Taiwanese language in the short time since her arrival. After 228, however, the Taiwanese around her treated her frostily.

When Mrs Zhao shared her memories of 228 with Fahey, it had already been forty-five years since the killings, and martial law had been over for five years. Yet, she was still cautious.

'She told me that it had been a very dangerous thing to discuss even in private for many decades after that, but she thought it was OK then,' Fahey said. 'Besides, she said with a laugh, "What can they do to such an old woman like me now?"'

In 1900, the Japanese founded Taihoku New Park in the centre of the administrative capital of its new colony. Today, the park – home to the radio station-turned-museum that broadcast the call for the Taiwanese to rise up against the KMT – is known as 228 Peace Memorial Park. Ever since the late 1990s, it has been a memorial to the victims of 228. While many people in Taiwan would prefer to never think about 228 and its ramifications, it remains an incendiary topic for many Taiwanese today.

On 28 February 2023, the new mayor of Taipei, Chiang Wan-an, stood in a sombre dark suit to deliver the annual apology speech outside the park's monument to the victims of 228. A former Silicon Valley lawyer, the rising KMT star claims to be Chiang Kai-shek's great-grandson via an illegitimate bloodline. Before he delivered his speech, young Taiwanese protestors, some holding banners, others shouting, objected to the heir to the Chiang legacy addressing the event.

At Chiang began to speak, the stage was rushed by more than a dozen protestors – none of whom made it to the podium. Security staff restrained the angry youth, some of whom called on Chiang to kneel and apologize to the Taiwanese people. At the centre of the chaotic scene, Chiang acted as if nothing was happening around him, the corners of his mouth betraying what appeared to be a slight grin. Once the protestors had been cleared, Chiang spoke, without mentioning his family's connection to the suffering he was there to commemorate.[28]

For many Taiwanese people, 228 is the KMT's original sin and a major reason they can never trust the party. If 228 was the Chinese Nationalist Party's opening act in Taiwan, it would soon follow up with an equally ghastly encore: the White Terror.

CHAPTER 5

White Terror in Free China

On 1 October 1949, following the Communist conquest of most of northern China, Mao Zedong announced the establishment of the PRC atop the Gate of Heavenly Peace, or Tiananmen, to hundreds of thousands in the square below of the same name. Weeks later, an embattled Chiang Kai-shek and the remnants of his KMT government retreated to Taiwan, with the island becoming the last bastion of the ROC.

The flight to Taiwan took place early in the morning of 10 December 1949, when Chiang and his son, Chiang Ching-kuo, boarded the *Mei-ling*, a plane gifted by the US government to the ROC government. Named after Chiang's wife, the *Mei-ling* flew out of the south-western Chinese city of Chengdu, which was under siege by the Communists. After a brief stop in Guangdong Province, still under KMT control, father and son flew onwards to Taiwan. Their arrival marked the shift of the 'provisional capital' of the ROC from Chengdu to Taipei. Despite vowing to retake the country they had lost, the Chiangs and the ROC would never return to China.

Among the roughly two million refugees who crossed the strait from China between 1948 and 1950, including soldiers, their families and conscripted children, Chiang Kai-shek undoubtedly had the easiest crossing and the best prospects awaiting him in

Taiwan. He took residence in a Japanese villa on Tsao Shan, or Grass Mountain, in Taiwan's north, far above the burgeoning refugee populations in Keelung and Taipei below. It was a mountain retreat fit for an emperor.

Just like his first governor of the island, Chen Yi, Chiang Kai-shek underestimated the Taiwanese people and their capacity to see through blatantly patronizing gestures. In an effort to stabilize KMT authority in Taiwan, Chiang offered the Taiwanese a sacrifice: Chen Yi himself. Following the bloodletting of 228 in 1947, Chiang had recalled the reviled Chen to China, where he would serve as a senior adviser to the central government. In June 1948, Chiang named him the new chairman of Zhejiang Province – where Chen's (and Chiang's) ancestors were buried and where a larger population offered more opportunities for corruption. It was a promotion, but one that would set the stage for Chen's downfall.

As the inevitability of Communist victory in southern China became apparent, Chen attempted to defect. When Chiang learned of Chen's plan, he had Chen arrested and sent to Taiwan in April 1950, where he was imprisoned in Keelung. Two months later, on 18 June, Chen was executed at Machangding – the same riverside killing ground where three years earlier his vengeance had stained the water red with Taiwanese blood.

Chiang ordered the execution to be a grand celebration. Fireworks were doled out to Taiwanese, many of whom must have been genuinely happy to see the architect of the 228 bloodshed snuffed out. It was not lost on the Taiwanese, however, that in late March 1947, after the majority of the 228 killings, Chiang and his wife had made an inspection trip to Taiwan where they congratulated Chen for a 'job well done'.[1]

By the time of Chen Yi's execution, Taiwan had been under martial law since 19 May 1949 – just over one year. The declaration of martial law is commonly accepted as the starting point of what would become known as the White Terror, a period of thirty-eight years during which at least 3,000 Taiwanese and Chinese were executed and more than 140,000 were imprisoned for alleged political dissent against the ROC. Chiang Kai-shek's government incorporated the Japanese-built Taihoku Prison and other legacy edifices into a sprawling gulag of intimidation, interrogation, torture, imprisonment and execution, while also adding new purpose-built facilities. The White Terror corroded trust within Taiwanese society, fostering a culture of surveillance, self-censorship and paranoia.

Over four decades, the White Terror would gradually transform from overt military suppression under Chiang Kai-shek to more insidious methods of control, often involving secret police or KMT-affiliated triads under his son, Chiang Ching-kuo. In response, the Taiwanese populace gradually shifted from fear to defiance of Chiang Ching-kuo's regime. This defiance led to the founding of Taiwan's first opposition party, the Democratic Progressive Party (DPP), in 1986, contravening a law that forbade opposition parties. An unwell Chiang Ching-Kuo would end martial law the following year, in 1987, one year before his death.

Much of what the KMT did to its victims during the White Terror was documented at the time but languishes in poorly maintained archives. Rehabilitated as an opposition party in a democratic Taiwan today, the KMT has little incentive to open up the archives and remind the world of its sins. DPP-driven transitional justice in Taiwan has focused more on recognizing the victims than bringing the perpetrators – many of whom are still alive – to justice. Meanwhile, the KMT, which as a party has never apologized for 228 or the White Terror, has been relatively successful at framing the public reckoning with the past as a partisan move by a power-hungry DPP. For its part, the DPP is also aware that many young voters are more interested in their economic prospects than they are in reopening old wounds. Such an attitude, while pragmatic and understandable, slows down the process of reconciling with the past, with ramifications for the present and future.

Spanning four decades, the White Terror deeply scarred Taiwanese society and rendered later generations unaware of their family members' traumas and tragedies. Much remains to be uncovered today to bring closure and some modicum of comfort to countless families.

Chang Yi-jung's quest for the truth about her grandfather, Huang Wen-kung, took time to gather momentum. When she was growing up, her family barely spoke of him. However, in the late 1990s, when Chang was in high school, she read a manuscript that her uncle was writing that mentioned his father – her grandfather – who, he wrote, had been executed by the KMT.

When Chang asked her mother, Huang Chun-lan, what had happened to her grandfather in 1953, she was met with a wall of silence. Huang had never met her father. He had been arrested before her birth and shot dead by a KMT firing squad shortly after she was born.

Later, in her twenties, Chang came across a passage describing her grandfather's arrest while reading a history of 228 and the White Terror. Afterwards, a 2007 government exhibit on Chiang Kai-shek happened to include a document that changed Huang Wen-kung's original fifteen-year prison sentence to a death sentence. The order was written by Chiang Kai-shek himself.

Unable to let the story of her unknown grandfather go, Chang Yi-jung decided to face the ROC bureaucracy head-on. After making requests to the ROC central archives, Chang eventually received a stack of hundreds of documents related to her grandfather in 2008. Among these documents, Chang discovered copies of five letters that her grandfather had written.

One was addressed to her unborn mother, Huang Chun-lan:

Dear Chun-lan, I was arrested when you were still in your mother's womb. What a pity that we, father and daughter, can never meet! What could be more tragic than that? Although I have never seen you, held you or kissed you, I love and care for you just the same. I am so sorry that I cannot do my duty as a father, Chun-lan! Can you forgive your poor old dad?[2]

It took three years for Chang to successfully sue the government for the original letter, which she then gave to her mother. Years after first reading it, the letter still had a profound impact on Chun-lan.

'Every time she read it, it's the same,' Chang said. 'From the first word she starts crying. She had never seen her father, so it was like he didn't exist, but when she saw the letter she knew she had a father – and that he loved her.'[3]

Before execution, the KMT would allow prisoners to write farewell letters to family members they would never see again – or in some cases, children they would never meet. Hundreds of these letters have been discovered within the poorly maintained ROC archives, giving brief, painful glimpses into the personal tragedies inflicted upon Taiwanese society by the KMT. The majority of them will never be delivered to their intended recipients, many of whom passed away long ago.

In one letter, Liu Yao-ting apologized to his wife for his 1954 execution:

My Yueh-hsia, you must listen to what I have to say. Even though we are apart, our hearts are connected. I deeply hope you can conquer all hardships, be brave, and not be heartbroken and lose health because of me.

Yueh-hsia, I'm sorry. It should be me who is looking after you and the children. This is also what I hope the most for the future. But at this moment I am incapable of doing this. Yueh-hsia, I hope deeply that you can forgive me.[4]

The White Terror affected everyone in Taiwan, including the Indigenous Taiwanese. Writing to his wife in 1954, Uongu Yat-auyongana expressed a defiant and protective instinct for his family's assets against potential KMT seizure:

> Chun-fang whom I long for. How glad I am to know you are healthy and well. 'No amount of gold, silver or gems are more precious than our darling children.' Do you remember the song? As long as we have our homes and our land we will be fine, because there are so many amazing children. No matter if our possessions are confiscated, my innocence will be revealed . . .
> In the fields and in the mountains, my spirit will always be. Don't give up the land![5]

The White Terror would last thirty-eight years. When it ended in 1987, it was the longest period of martial law the world had ever seen – an ignominious record that would later be broken by Syria. In a way, this era of martial law and the shared suffering of the Taiwanese people would create the foundations for a new Taiwanese identity decades later – the White Terror targeted every group in Taiwan: for the *waishengren* or mainlanders who had come with the KMT from China, it was an endless anti-Communist witch hunt; for the *benshengren* or native Taiwanese, it suppressed calls for self-determination and Taiwanese independence.

Between 1949 and the late 1980s, Taipei served as the main processing centre for the political persecution of the people in Taiwan, targeting both native Taiwanese and mainlanders. The system didn't discriminate based on ethnic or political backgrounds, gender or age – any and all threats to the system were to be rooted out, processed and, eventually, imprisoned or executed.

The military is the foundation of martial law and, as martial law is usually unpopular, tight control of the officers and soldiers who impose your will is paramount in remaining in power. To this end, Chiang Kai-shek founded the Fu Hsing Kang College in 1951, north of the city across the Keelung River. Fu Hsing Kang graduates would serve as political officers – thought police – within the military. These thought police pervaded Taiwanese society, and were even stationed in Taiwan's schools, making middle and high schools more reminiscent of military academies than institutions of learning and development. Political officers would ensure that teachers adhered strictly to the approved curriculum, which taught that everyone in Taiwan is Chinese, and that the ROC is the one true Chinese government and would soon retake the mainland.

Students were forced to learn Chinese geography, history and culture, with no space provided for Taiwanese history, culture or identity. This curriculum introduced the politically loaded term 'mainland' to Taiwanese, who had previously referred to China as 'Tangshan' – a distinct entity to their homeland. Favoured today by both the KMT and the CCP, the term 'mainland' implies that Taiwan is simply an offshore part of China. The phrase 'mainland China' is also frequently repeated by international media, academics and others when discussing Taiwan, suggesting they either do not understand or else do not think about its implication (or, perhaps agree with that implication).

Despite present-day Taiwan's accolades for being a bastion of freedom and democracy, political officers are still stationed in schools. Their role, however, has changed, alongside Taiwan's curriculum, which has increasingly highlighted Taiwanese history and identity. Serving mostly as helpers on school campuses, the snoopiest thing today's political officers do is make sure kids aren't smoking cigarettes outside of school. But their presence is a reminder of the lingering aspects of martial law that can still be found in Taiwanese society. Taiwanese call these relics of their dark past *yidu* (遺毒), or 'residual poison'.

The White Terror's machinery was merciless. After the system identified you as a problem, your ordeal would typically commence with a loud, unexpected knock on your door by white-helmeted military police. The secret police working under Chiang Ching-kuo enjoyed carte blanche to arrest anyone they wanted, at any time, for any perceived transgression against the Chiang regime and the KMT party-state. The next steps were almost always interrogation, sentencing and punishment.

Once taken into police custody, victims were subjected to physical and mental torture. Between interrogation and sentencing, detainees were moved into the prisons of the KMT secret police. There, they would be crammed into hot, overcrowded cells that stunk of body odour, urine and faeces, with prisoners often forced to sleep while squatting or standing. Inmates were deprived of exercise and sunlight, leading to rapid physical and mental deterioration. The sites of many of the most dreaded addresses in Taipei are now popular tourist destinations. Ximending, a bustling neighbourhood that is popular with young people and the LGBTQ+ community, was home to the Intelligence Section headquarters. The building was a temple during Japanese rule, but under the KMT, it was a prison that became renowned for cruelty. The Sheraton Grand Hotel near Taipei Main Station was built on the site of a former Japanese-era army warehouse that was converted into the Martial Law Section Jail. Hidden behind high concrete walls topped with barbed wire, the wooden structure was constantly damp, filthy and putrid, with as many as thirty detainees packed in a twenty-square-metre cell. The Ningxia night market in Taipei is one of the better night markets in the city today, frequented by both local Taiwanese and foreign tourists. Just across the street from the market is the former site of the Provincial Police Headquarters Detention Centre. There, squat-style toilets in the floor were the only source of water for brushing one's teeth or washing. Green Island in Taitung County, roughly 30 kilometres off Taiwan's south-eastern coast, attracts coral-seeking snorkellers today. From 1951 to 1965, however, it was home to the New Life Correction Centre. Here, prisoners were forced to reform themselves through back-breaking labour. Donghe Township, also in Taitung County, is known for having some of Taiwan's best surfing and for its *baozi*, Chinese-style steamed buns filled with meat, vegetables, or sweet red bean or black sesame paste. During the White Terror – when beaches became restricted military areas and surfing was illegal – it was home to the infamous Taiyuan Prison, officially known as the Taiyuan Vocational Training Centre. Following its completion in 1962, it replaced Green Island as the largest facility for the incarceration of political prisoners in Taiwan, who endured force labour during their stays.

Sentencing was the most straightforward part of the political prisoner's ordeal and was carried out in military courts. Judges were mainly there to provide a veneer of legitimacy to state violence and intimidation. The so-called justice that was meted out was solely determined by the KMT. There were two types of punishment: execution or imprisonment. Those who were sentenced to death were taken directly to an execution ground and shot. Two main locations in Taipei were used for this, both located on the Xindian River. One was the site of the Japanese-era horse racing track at Machangding, where Chen Yi faced the firing squad. The other was further upstream at the Ankeng Execution Ground. The names of the executed would be posted on a notice board daily outside Taipei's main railway station. During the early days of the White Terror, the names were also published in newspapers.

To the south of Taipei, the stadium-sized Jingmei Detention Centre was home to secret military tribunals that sentenced at least 1,100 victims to execution. For those who were spared a quick death, the compound also housed a prison that could hold hundreds of prisoners. As elsewhere, inmates were subjected to incarceration under inhumane conditions, and tortured.

Former political prisoner Roger Hsieh served two sentences in Jingmei during the 1960s and 1970s, both connected to his involvement in helping his professor, Peng Ming-min, prepare leaflets advocating the overthrow of Chiang Kai-shek. In his book, *Talking About the Martial Law Section Jail in Jingmei*, Hsieh recounted the kinds of torture to which he or his fellow inmates were subjected.

Among the forms of torture that Hsieh documented were the plucking of fingernails, a pointed stick rammed into the anus, petrol squirted up the nose, force-feeding of limes, electric shocks, forced drinking of urine, genital torture, forced eating of dog faeces, beatings of prisoners while hanging from the rafters by both hands or feet, the removal of teeth without anaesthesia and the forcing of inmates to listen to the screams of others as they were relentlessly tortured.[6] Many of those who were sentenced to political imprisonment were transported to a military stockade, where they would wait to be shipped off to their new homes.

Despite the overwhelming power and cruelty the KMT wielded during the White Terror, the Taiwanese people continued to resist, regardless of the great personal risk involved. Many of the Taiwanese who lived abroad or who had managed to flee before the declaration of martial law became part of a diasporic resistance that, decades later, would be crucial in the transition from martial law to democracy. Overseas resistance to the KMT began in a post-war Japan that was still rebuilding, before eventually shifting its centre of gravity to the US.

One of the resistors was Chen Chih-hsiung, who was born in 1916 in what is now Pingtung County in Taiwan's southwest. Thanks to his proficiency in Dutch, Malay, Japanese, English, Taiwanese and Mandarin, Chen became an indispensable asset to the Imperial Japanese Army when it occupied Indonesia in the 1940s. Following Japan's surrender in 1945, Chen chose to stay in Indonesia, becoming an integral part of the struggle for independence from Dutch colonial rule – a struggle that landed Chen in prison for a year. Chen's dedication to Indonesian independence caught the attention of the Indonesian nationalist leader Sukarno who, upon the success of the revolution, honoured Chen with Indonesian citizenship. This relationship proved pivotal for the budding Taiwan independence movement led by Thomas Liao, a Japan-based activist who had founded the Taiwan Democratic Independence Party in Kyoto. In the pursuit of this cause, Liao also founded a provisional Republic of Taiwan government-in-exile on 28 February 1955.[7] The choice of date (228) was no coincidence, and was one that would later be favoured by future Taiwan independence activists for demonstrations and actions.

Chen Chih-hsiung's move to Japan led him to support Liao's independence activities, including helping Liao attend the pivotal Bandung Conference hosted by Sukarno in 1955, where leaders of Asian and African countries convened. Liao appointed Chen as the Republic of Taiwan's ambassador to Southeast Asia, a role that, despite his honorary Indonesian citizenship, eventually made him the target of the Indonesian government. He was arrested, had his passport rescinded and was deported as a result of Sukarno's new friendship with Beijing. Now stateless and seeking refuge, Chen moved to Switzerland before returning to Japan, only to be kidnapped by KMT agents and forcibly returned to Taiwan. After being forced by the KMT to renounce his independence activities, Chen eventually founded another anti-KMT party in 1961 and was arrested the following year, before being executed in May 1963.

In 2013, fifty years after Chen's execution, his daughter, Vonny Chen, finally learned why the KMT killed her father. Vonny Chen had grown up in Indonesia with the family her father had left behind in order to fight for Taiwan. The ROC government had only informed Vonny of her father's death in 1979 – sixteen years after executing him. On one of several trips in the 1980s, an aunt in Taiwan's Yilan County confirmed to Vonny and her two brothers that their father had been executed. The aunt warned them not to ask about why he was dead.[8]

During her 2013 visit, Vonny Chen was given a stack of documents related to her father from the central archives. Among them was a will in which he asked a family friend to help take care of his three fatherless children. Chen lamented that had her father's friend seen the will, her and her brothers' difficult lives could have been much better. She also received the final letter written by her father, which was addressed to his children and written in Japanese. 'I died for the Taiwanese people,' the elder Chen wrote.

While in Taiwan, Vonny Chen met with one of her father's cellmates, who described the conditions of his execution. Guards stuffed Chen's mouth with a rag, he told her, so that he couldn't yell 'Long live Taiwanese independence!', as he had upon his arrest. They also hacked off both of his feet so that he could not stand with dignity when they shot him.[9] Chen Chih-hsiung became the first person executed for the cause of an independent Taiwanese state. One might think that would make him a hero in today's democratic Taiwan, but few Taiwanese know his name.

'Some say my father is a hero,' Vonny Chen told the *Taipei Times* during her 2013 visit. 'But for Taiwan, he's not a hero. That's why they killed him. My father would be Taiwan's hero if it were independent. But if not, nobody knows about him.'

Seven years after Chen's execution, Taiyuan Prison in eastern Taiwan was the site of one of the only armed uprisings against the KMT during martial law. In February 1970, five inmates attacked six guards during a shift change, wresting their guns from them and stabbing one to death. The inmates had planned to pilfer guns and ammo from the prison's storeroom and then – as in the early moments of 228 – overtake a local radio station and incite an uprising against the KMT. Their plans were foiled when other guards overheard the initial struggle. The five inmates fled into the surrounding hills but all were eventually captured and executed.

Even though the rebellion was quickly quashed, it signified a rare and direct challenge to KMT martial law. It was widely suspected that many of the inmates and guards were sympathetic to the failed uprising. This rattled the Chiang regime, which responded by building a new maximum-security prison on Green Island which lacked any of the 'perks' of its predecessor. It was cruelly named Oasis Villa.

'Free China'

The failed Taiyuan Prison uprising aside, the vast majority of those imprisoned or executed by the KMT party-state were not seeking to overthrow the Chiangs – although it would be hard to blame anyone who was fed up with their rule. They were simply demanding relief from the oppression of martial law in their homeland. The party-state had perversely framed Taiwan as the 'Free Area of the Republic of China', or 'Free China', even as it stifled the freedoms of its own citizens. Among the anti-communist bloc, especially in the US, the excesses of ROC rule in Taiwan would be justified as a necessary sacrifice in a much bigger struggle.

The ROC's flight to Taiwan marked a significant transformation not only geographically, but also in terms of its governance structure. Once in Taiwan, the ROC effectively became a rump state, only retaining control over Taiwan and a few islands near the Fujian coast. As a result, power largely moved away from the central government, which dealt mostly with foreign relations and national defence. The Taiwan provincial government replaced the central government in terms of importance and influence over the day-to-day running of the country. In a 1955 report from Taiwan, *Time* magazine noted:

> The national government has been progressively diminished as the provincial government of Formosa has increased its independence, until

today there are only 12,000 employees in the national government v. 113,000 in the provincial government. Except for Foreign Affairs and the Defense Ministry, most of the national ministries, their functions duplicated by provincial departments, are only skeleton organizations with nothing to do but plan for the day of The Return.[10]

With considerable propaganda assistance from Washington (which prioritized the Chiangs' role in the fight to contain communism), the KMT was largely successful at getting the world to see the ROC in Taiwan as 'Free China', where life was normal and only a fringe handful of bad elements – painted as being Communist-aligned – opposed the regime.

From the very beginnings of the KMT's relocation to Taiwan, life for Taiwanese or Chinese was anything but free. The freedoms and rights of the populace were severely curtailed under the guise of preventing Communist infiltration. Chiang Kai-shek and the KMT had lost China to Mao and the Communists, and they certainly weren't going to lose Taiwan to the Taiwanese, to whom they felt vastly superior. A new secret police force was formed in August 1949, with Chiang Ching-kuo in charge. Chiang Kai-shek's reliance on his son for such a crucial role underscored the lack of trust he had in those who had fled China with him. To be fair to the Generalissimo, this was understandable, given the large number of KMT generals and other officers who had defected to the Communists as his control over China crumbled.

Chiang didn't like dealing with people, but he clearly thought he deserved to rule them. That same *Time* report from 1955 described Chiang's morning routine. After rising before 6 a.m., he would pray and meditate for one hour, sometimes joined by his wife, Soong Mei-ling. 'It is then that he gets his strength for the day,' she told *Time*. Eventually, Chiang would descend Yangming Mountain, as Grass Mountain had become known. Chiang renamed the mountain after Wang Yang-ming, the Generalissimo's favourite neo-Confucian thinker and philosopher of the Ming Dynasty, who hailed from his home province of Zhejiang in China.

'After breakfast and a careful scanning of Formosa papers and others flown in from Hong Kong, Chiang dons his khaki cape, enters his 1949 Cadillac, and makes the 25-minute drive to his office in the Ministry of National Defense in downtown Taipei,' *Time* wrote. 'Soldiers of the security force appear as if by magic along the route, then as magically melt away after he has passed. Past a dark bronze bust of himself on the stair landing, he walks quickly and alone to his third-floor office, where the blue velvet curtains are always drawn for security.'

Outside Chiang's court, the White Terror had been gathering momentum, encroaching more and more on the daily life of the average person. In 1950, the KMT instituted a ban on overseas tourism or family visits without special government permission. The ban would last until 1979. A 1952 law saw the government determine how long middle school students' hair could be: one centimetre at longest for boys and nothing below the ears for girls.

Media was brought under tight government control in 1953, with laws mandating that newspapers and magazines cannot malign the head of state or national policy, along with vague rules against confusing society or sowing divisions – rules that could be interpreted in any way that suited the KMT. That same year, government censors investigated school textbooks, removing any books from circulation that featured implicit criticism of the government or discussion of class struggle.

The KMT's security architecture was given an overhaul in 1958, with the founding of the Taiwan Garrison Command, which oversaw everything from prisons to censoring news items. One decade into KMT rule, the Taiwan Garrison Command informed the government that it had removed the names of more than 125,000 people from the citizen registry. Their removal, it was explained, was due to the Command being unable to find them. Given the high degree of control the KMT police state had over civilians and their movements in and out of Taiwan, this figure suggests a significant toll of imprisonment, disappearances and executions beyond the already known atrocities of the 228 massacres and the early White Terror period. It also likely includes no small number of people who had gone into hiding within Taiwan.[11]

Amid this climate of fear and repression, Chiang attempted to instil a sense of Chinese nationalism and cultural continuity – whether people wanted it or not. The ROC's exodus to Taiwan was a deeply traumatic chapter for Chiang, but also for those who fled with him. These mainlanders, or *waishengren*, were a crucial base of support, administering most of the ROC apparatus. For these people and himself, Chiang attempted to recreate a semblance of China in Taiwan – a Taiwan that was still very Japanese in terms of its recent history and identity.

In asserting his claim to rule 'Free China' in Taiwan, Chiang ordered the construction of a new Palace Museum in the hills north of Taipei, aiming to symbolize the continuity of Chinese cultural identity. In preparing for his departure from China, Chiang Kai-shek ordered an unknown quantity of gold, estimated to be millions of taels, to be shipped from Shanghai to Taiwan. Chiang Ching-kuo oversaw the top-secret mission which would provide a financial safeguard for the refugee government. Gold wasn't the only prize that made the journey across the Taiwan Strait, beyond the reach of the Communists. Over 600,000 priceless artefacts from the Chinese emperors' personal collections were eventually transferred from the Palace Museum in Beijing to Taiwan.[12] Architecturally, the Palace Museum in Taiwan museum is a mix of Chinese and brutalist architecture, with traditional white *paifang* arches out front, a green-tiled roof with orange trim and a sprawling, white-tiled edifice built into a lush hillside. Since the museum opened in 1965, it has been considered to have the best collection of Chinese art and artefacts anywhere in the world.

The museum stands as a landmark of the ROC evacuation to Taiwan. It amplified the KMT's claim to be the protectors of Chinese culture, a claim that gained particular significance during Mao's Cultural Revolution. During this period, countless cultural treasures were destroyed by Red Guards and other Chinese caught up in the ideological fervour in a decade-long bid to purge China of its historical legacy. Hence, the museum bolstered Chiang's claim to be the legitimate ruler of all of China, but also played a crucial role in imposing a Chinese identity on Taiwan. The message to the Taiwanese was clear: you are Chinese now.

Nowadays, Chinese nationalists and their online supporters often accuse Taiwan of stealing the Chinese treasures and gold that the KMT made off with. In response, Taiwan nativists often reply that they'd be happy to return it all in exchange for China renouncing its claim on Taiwan.

The construction of an alternate China – one dramatically different from the idealized New China promised by US propaganda leaflets in 1945 – extended into almost every aspect of daily life. This included time itself, which was reoriented with years counted according to the founding of the ROC in 1911, when Taiwan was part of Japan, and marked time according to Japanese reign names. Some older Taiwanese people today still think of events during the pre-democracy days primarily according to the Republican calendar, while for younger people, the two sets of time are more or less interchangeable. In conversations I've had, Taiwanese seniors seem more capable of placing my birth year – 1976 – in the context of their own life if I convert it to year 65 of the Republic. Even receipts still use the ROC calendar. If you were to buy an iced tea and a rice triangle at a convenience store in 2025, your receipt would show the same month as the Gregorian calendar, but the year would be 114.

In the process of rewriting Taiwan's identity, streets were renamed after Chinese cities. In central Taipei, an overlay of a map of China guided the renaming of streets. Streets in the capital's north-east were named after north-eastern Chinese cities or provinces, such as Changchun or Liaoning, while in the south-west, roads were named after south-western Chinese cities, such as Chengdu, Kunming and Guilin. This renaming affected countless villages and townships across the island, with new road names often embodying traditional Confucian values such as *Ren'ai* (benevolence), *Zhongxiao* (loyalty and filiality) and *Xinyi* (good faith). Sun Yat-sen, the founding father of the ROC, received significant honours. His name, usually rendered in Mandarin as Sun Zhongshan, was given to major roads, urban districts, primary schools and even a university in Kaohsiung. In Taipei, Roosevelt Road was named in recognition of Franklin D. Roosevelt's role in the Cairo Declaration, which paved the way for the ROC's governance over Taiwan. This major thoroughfare connects southern Taipei to the city centre, where it becomes Zhongshan Road, yet another nod to Sun Yat-sen.

Part and parcel with the Sinification of Taiwan was the pervasive influence of 'Chiangification' – the elevation of Chiang Kai-shek and his values above all else. Even today, his adopted name, Zhongzheng, is used for roads, schools and city districts. Streets still bear the names of the virtues he extolled, such as the New Life (*Xinsheng*) movement that aimed to stop the Communists with moral fortitude, or the Four Cardinal Principles (*Siwei*) and Eight Virtues (*Bade*) of Confucianism. Statues of Chiang were erected throughout Taiwan, in its schools, military facilities, parks and other spaces. Today, the statues are officially viewed as authoritarian symbols and are gradually being relocated to a park outside Chiang's mausoleum in Taoyuan County. This park has taken on a unique character, hosting an array of differently coloured statues of Chiang (along with a few of Chiang Ching-kuo and Sun Yat-sen) arranged in ways that feel like they're interacting with each other – some seated, some standing, some making proclamations, others on horseback.

The era of Chiang Kai-shek and the subsequent years under the rule of the KMT were defined by a deep-seated cult of personality surrounding Chiang himself. This period saw the mobilization of both the ROC's resources and the KMT's influence to ensure Chiang remained beyond public criticism. Educational curriculums were tailored to instil reverence for Chiang. Children were forced to learn his sayings, while older students were forced to study his essays. In many ways, the near-deification of Chiang parallels the present-day efforts of the CCP to inculcate Chinese people with a similar reverence for Xi Jinping.

Both domestically within the ROC and among its anti-Communist allies, Chiang Kai-shek was lauded as a benevolent, principled leader committed to halting the spread of communism. However, this image contrasted sharply with the realities recognized by foreigners residing in Taiwan – especially the Americans who were working with the KMT. In a 1999 interview, the retired American diplomat Harvey Feldman spoke about the Chiangs' rule in Taiwan, where he was stationed across the 1960s and 1970s.

'The Chiangs ran a pretty tight ship,' Feldman recalled. 'The press was totally fettered; all the media was captive. There were only three TV stations: one was owned by the national government; one was owned by the provincial government; and one was owned by the Chinese Nationalist Party [KMT]. There was press censorship. Taiwan had all of the attributes of an authoritarian martial law state. It may not have been as harsh as the regime on the mainland. A friend of mine described the Taiwan situation as "soft totalitarianism".'[13]

One of the more famous examples of the regime's paranoia and intolerance is the case of Bo Yang, a well-known cultural critic and translator. A translation of the American comic strip *Popeye*, of all things, would be Bo's undoing. In the comic, Popeye and his son find themselves on a small island, where he runs for president. Bo translated Popeye's opening words to voters as 'Compatriots of our country . . .', echoing the words Chiang Kai-shek used to address

Taiwan. He later wrote that he had no ill intent with his translation. On the night of 4 March 1968, the military police showed up at his home. Bo was tortured until he made a false confession of being a member of the CCP. The Popeye comic strip seems to have been the last straw for the KMT, who had tired of Bo's often negative portrayal of the party in his writing. Bo was sentenced to twelve years at the Jingmei Detention Centre for spying for Beijing and 'attacking the country's central leadership'. Good news came in 1975, when Chiang Ching-kuo succeeded his late father and announced the reduction of sentences for political prisoners by one third, which meant that Bo should have been released in 1976. Instead, he was put under house arrest on the island, which he believed was due to the elder Chiang's contempt for him. Bo was eventually released from Green Island on 1 April 1977, more than a year after his reduced sentence ended. His freedom came following pressure on Chiang Ching-kuo's government from Amnesty International as well as US Congressman Lester Wolff. Bo later founded the Taiwan chapter of Amnesty International in 1994 as the country was democratizing.

Not everyone was as fortunate as Bo Yang. In 1970, a businessman on a work trip to the British colony of Hong Kong, Chen Chen-hsiung, wrote a letter to Chen Yi – not the executed former governor of Taiwan, but a vice-premier of Communist China – asking that China not invade Taiwan. After the ROC government learned of the letter, he was tried in secret by a military tribunal. Despite his letter not being supportive of communism or Beijing, Chen was executed in November 1970 for the crime of attempting to contact China's government. He was thirty years old.[14]

Even Taiwanese citizens abroad could not escape the KMT's far-reaching influence during the White Terror. The KMT dispatched 'professional students' to countries where large numbers of Taiwanese were studying, especially the US. These student spies would report unapproved activities back to Taipei, often landing students on blacklists that prevented them from returning to Taiwan. The student spies helped maintain a climate of fear and control that extended beyond Taiwan's borders, specifically targeting those who opposed the KMT's authoritarian rule.

Tun-Hou Lee is Professor Emeritus at Harvard's T. H. Chan School of Public Health. Shortly after Lee moved to Boston on 4 July 1976, his consciousness as a Taiwanese – and ROC – citizen began to change.

'The turning point was my realization of the depth of various KMT brainwashing efforts and the existence of KMT spies on campus,' Lee told me. 'Boston was a kind of battleground between KMT student spies and operatives and Taiwanese students and immigrants who were determined to defeat the authoritarian rule of the KMT. There were student spies who monitored dormitory mailboxes to record who had received anti-KMT publications.'

In the late 1970s, one of the best-known 'professional students' in Boston, he said, was Ma Ying-jeou, who would go on to be both KMT chairman and ROC president and is still active in Taiwanese politics. Ma has been publicly accused of being a student spy in Boston many times over the years, which he has consistently denied.

'I despise Ma because he not only lacks the courage to acknowledge that what he did back then was wrong, but he still continues to accuse others of smearing him,' Lee told me.

For Taiwanese who emigrated to the US or studied there during the White Terror, America was a vast country with freedoms that were unimaginable back home. Yet it was also an inescapable fact that the self-professed land of the free played an instrumental role in keeping the Chiangs in power, while fully aware of their 'soft totalitarianism', as the diplomat Feldman noted.

The End of Martial Law

For most Taiwanese living in Taiwan towards the end of the White Terror, martial law was the natural state of affairs – it was all they had known. Transitioning away from this era was a complex and gradual process. Even after martial law was lifted in 1987, it took time for people to internalize the new freedoms they had. According to Maggie Lewis, a law professor at Seton Hall University, this gradual acceptance included those who were directly involved in martial law's dismantling.

'Taiwan's path out of martial law was more like turning an aircraft carrier than turning on a dime,' she told me. 'Vestiges of martial law lingered far after martial law was officially dropped.'

This was in part due to the gradual shifts in personnel, Lewis said. Career civil servants, including judges and prosecutors, remained in their posts during the multi-year transition.

'Even for the many who fully embraced leaving the authoritarian era behind, it was a transition to a system that had radically different procedures,' she noted.

The laws also took time to shift.

'When I first went to Taiwan in 2006, for example, it was to study a lingering law that allowed for a person's liberty to be denied for three years based on a secret witness system,' she said. The Constitutional Court gradually invalidated aspects of the law and, finally, the legislature – at the urging of the executive branch – did away with it.

The end of the White Terror also took a while to sink in with those Taiwanese born after 228. After living all one's life restrained by internalized red lines, it took many Taiwanese people years to realize that those lines had disappeared.

Towards the end of 2023, I spoke with film producer and Taipei native Lee Lieh, then sixty-five, who began her career in the entertainment industry as an actress during the final decade of martial law. Sipping a hot chocolate at an Australian cafe near Taipei 101, Lee wore a grey hoodie and skinny jeans, her auburn hair in a bob. She spoke with the wisdom and empathy of someone who has seen it all. In her childhood and early adulthood, living under KMT martial law didn't seem unusual to Lee – because she had nothing else to compare it to.

'I actually didn't really have any feelings about it when I was younger, because everyone was living the same reality,' she said. 'That's how the common people are here – if you've got something to wear and something to eat, then you're getting by. People don't care, especially when they're young. You would basically start to slowly know what's going on once you'd grown up a bit.'

For Lee, she started figuring out Taiwan was 'different' during the 1970s: 'You'd hear some news or read some articles, and you'd start to figure out "Oh, OK, we're not quite the same as some other countries."'

Growing up in Taiwan, there was no question that she lived in the ROC.

'My country was the Republic of China – this was something that went back to before I was born. Taiwan, for me at least, represents the ROC. The ROC is Taiwan, that's how I feel. But I'm not like those people who think that mainland China is the ROC too. The only thing the ROC has is Taiwan. It used to be framed as the "ROC on Taiwan", but I'm not having that – the ROC is Taiwan and Taiwan is the ROC.'

Working in showbiz in the 1980s, she and her colleagues knew that 'there were some things you couldn't shoot, some things you couldn't say. But because you live in this environment, you know what you can't shoot, what you can't say. When suddenly there were some things you could now shoot, some things you could now say or write – that feeling was actually quite exhilarating.'

After the end of martial law in 1987, people were still reluctant to air their thoughts, views or questions in public: 'How should I put it? You wouldn't really be inclined to speak out then, because you were used to not doing so. But eventually, you might start talking within small circles of friends. Then you started to realize that the number of people willing to talk was steadily increasing. It was a slow process, it didn't happen overnight.

'Our society had told us that there were many things we couldn't say, that we couldn't encounter, listen to, read or watch.' Things began to change around 1990, Lee told me: 'People began to dare to speak things aloud to others.'

KMT martial law and the risks taken by those who attempted to learn more about the outside world were the focus of the 2019 horror film *Detention*, which Lee's company produced. Set in a high school in which the ROC military exercises almost total control over student life, the film follows a secret club of teachers and students who read banned books. One such reading group is discovered by the military. The subsequent violent and lethal crackdown on the involved teachers and students unleashes a ghastly spectre resembling a KMT officer. Speaking only in ROC propaganda slogans, it hunts and kills those involved in the reading group.

Detention not only documents the historical repression experienced during the White Terror, but also addresses the censorship and risks faced by those who sought to live outside the scope of the KMT.

Now, as Hong Kong's present reality increasingly resembles Taiwan's past, many people from Hong Kong's film industry have relocated to Taiwan, Lee told me. Among them is the actor Chapman To, who was blacklisted from Chinese and Hong Kong films following his vocal support for the 2014 Sunflower Movement in Taiwan and the Umbrella Revolution in Hong Kong. He obtained citizenship in Taiwan in 2022.

'What's intriguing is that every one of [the Hong Kong creatives who have relocated], when writing a story, would ask me, "Is it OK for me to write this? Can I film this?"

'I always tell them, you can shoot or write anything in Taiwan. We don't have things you can or can't film or write, we only have things that are written or filmed well or poorly. But they don't really believe me.'

Lee said she was struck by the sensitive political 'radar' that Hong Kong actors had developed because of their government's assault on freedoms.

'When you see Hong Kong actors come to Taiwan and they're worried about this and worried about that, then you know – Hong Kong's White Terror has already begun.'

Even though Lee's company is unlikely to have its films screen in China in the foreseeable future – when *Detention* was released, the film's Chinese name, literally 'Return to School' was censored on China's internet – the advent of streaming has made it possible for Taiwanese media to reach a wider global audience than ever before, she said.

One might think the CCP would be more than happy to point out the human rights abuses of its erstwhile nemesis, the KMT. Given that the CCP and KMT are both Han ethnonationalist parties with Leninist roots, however, it makes sense that Beijing isn't seeking to let its citizenry look back on the past. Especially as, while Taiwan was under martial law, Mao unleashed tragedies including the Great Leap Forward and the subsequent famine that starved thirty million people to death.

That said, Lee still finds it 'hilarious' that works like *Detention*, which hold no connection to modern China whatsoever, are banned across the strait.

'This was a film about the KMT-era White Terror,' she said. 'So what are you all nervous about? It's got nothing to do with your country.'

Since 2017, Taiwan has celebrated its relatively recent freedom of speech on Freedom of Expression Day, which falls on 7 April. In 2019, I visited the Nylon Cheng Memorial Museum to learn more about the man whose death on 7 April 1989 the holiday commemorates. The museum is in a residential building tucked inside a narrow alley in Taipei's Songshan District that is nicknamed Liberty Lane. A wall running along the opposite side of the alley from the museum featured murals of human rights activists, including Nelson Mandela, Liu Xiaobo and Malala Yousafzai, with notable quotes of theirs in Chinese and English.

I rang the doorbell outside the old building and was buzzed in. I walked up to what was once the office of *Freedom Era Weekly*, a magazine that in the late 1980s advocated the overthrow of the ROC government and the establishment of a Taiwanese state.

Following the official end of martial law, the magazine's firebrand publisher and editor, Cheng Nan-jung – also known as Nylon Cheng – was a driving force behind calls for total freedom of speech, especially regarding the one subject the KMT was loath to allow: Taiwanese independence.

Nylon Cheng was born in Taipei in September 1947 to a Chinese father and Taiwanese mother. Cheng's mother was pregnant with him during the chaos of 228, and she was given shelter by Taiwanese neighbours during the initial period of Taiwanese revenge violence against *waishengren*. Cheng would later write about how much it affected him to think that even before he was born, his Taiwanese neighbours were protecting him. Both this and the repression of the White Terror instilled a deep sense of Taiwanese identity in Cheng during his youth.

Overthrowing the ROC was more than one journalist could do, so Cheng instead aimed to mobilize an increasingly vocal and restless Taiwanese citizenry to put an end to the KMT's censorship regime. *Freedom Era Weekly* was but one vehicle for that goal. Cheng had cleverly registered more than a dozen publication licences for similarly named magazines – what he jokingly called 'spare tyres' – that he would use whenever the government revoked the license he was using at the time.

In early 1989, less than two years after the end of martial law, Cheng was once again in the hot water that he was already very familiar with, but it had never been this hot. *Freedom Era Weekly* had crossed a KMT red line by publishing a draft constitution for a Republic of Taiwan. It was a blatant violation of the anti-sedition law that remained on the books following the end of martial law. Police were sent to the magazine's office to arrest Cheng, but he had barricaded himself inside, and told the cops that they would never take him alive. He wasn't joking.

After two months of waiting for Cheng to give himself up, on 7 April the commanding officer for the case, Hou Yu-ih, decided enough was enough. As the police began to break through the barricade, Nylon Cheng lit his office, and himself, on fire. A television crew was there on the scene, broadcasting live as smoke billowed out of the third floor. The fire department eventually extinguished the conflagration, and when police entered, they found Cheng's charred corpse on the floor, burned beyond recognition, his hands held in front of him, curling inwards unnaturally, his mouth agape. He was forty-one years old.

The Taiwanese public was shocked at his death, its anger further stoked by the police crackdown on a funeral procession for Cheng weeks later, in front of the Presidential Office. When the unauthorized funeral march arrived in front of the presidential building, the police were waiting, and had placed coils of razor wire across the street. Shortly after police began blasting the mourners with water cannon, one of the marchers, thirty-two-year-old activist Chan I-hua, doused himself with petrol that he had been carrying, set himself ablaze and threw himself onto the razor wire barricade. Chan stood up one last time, lit up like a torch but appearing at peace, as he looked towards the photographer Pan Hsiao-hsia, who captured his transformation from activist to martyr.[15]

Cheng and Chan's self-immolations were ultimately successful in generating public pressure on the KMT, which went on to repeal its anti-sedition laws. In 2024, Reporters Without Borders – whose Asia headquarters is in Taipei – rated Taiwan the world's twenty-seventh freest media environment, putting it ahead of nearby democracies including Australia, South Korea and Japan.

The small museum's walls were filled with photos of Cheng, many of which showed him clutching a mic with his left hand and pumping his right fist in the air as he energetically addressed a crowd, his trademark yellow pro-democracy headband reading 'New Country Movement' in Mandarin tied over his mop of black hair.

The day of my visit to the museum was the thirtieth anniversary of Cheng's death. One of the other few people in the museum was Yeh Chu-lan, a former deputy premier under President Chen Shui-bian and a senior adviser to Tsai Ing-wen. We chatted briefly about how far Taiwan's freedom of expression had come since Cheng's fiery death.

One photo showed Cheng being escorted in handcuffs by two soldiers, a smile on his face as he raised his cuffed hands almost triumphantly at the photographer who recorded the moment.

'Look at that smile,' Yeh said. 'There was nothing they could do to stop him.'

Nobody knew that better than Yeh – Cheng Nan-jung was her husband.

Just over her shoulder was a photo of a forty-year-old Yeh in that very room in 1989, looking down at her husband's blackened corpse. Three men, one a Presbyterian minister, stood next to her as she processed her loss and the excruciating pain of her husband's last moments of life. One of the men had his hand on her shoulder, consoling the new widow who would have to raise her nine-year-old daughter, Cheng Chu-mei, on her own.

Today, one room in the museum maintains the burned remnants of Nylon Cheng's personal office – the room in which he was found – as a reminder of his sacrifice. Cheng Chu-mei, who grew up fatherless, is now the chair of the Cheng Nan-jung Liberty Foundation, which along with the museum is also based in the former magazine office.

Yeh returned to the museum on Freedom of Expression Day in 2024, thirty-five years after she lost her husband to his ideals. A reporter with the Central News Agency (which was once a KMT propaganda outlet) asked her what she would say to her martyred husband.

'I'd say that you went too far by leaving me and our daughter so soon. But I'd also say that if I had another chance, I would give anything to fall in love with you again,' Yeh said. 'I'd also say thank you, because if not for you, Taiwan wouldn't be this good now.'[16]

CHAPTER 6

Death of a Dynasty

Throughout the 1930s and 1940s, Chiang Kai-shek navigated a tumultuous era in China, marked by a brutal eight-year-long war with Japan, his kidnapping and near-execution by one of his generals, and the bloody revolution led by Mao Zedong. In Taiwan in the 1950s and 1960s, he kept an iron grip on power. But he also enacted land reforms that unleashed the entrepreneurialism of the Taiwanese people, launching the four decades of rapid economic growth that would become known as the Taiwan Miracle. His empire had shrunk, but at least he still had one.

The 1970s would not be kind to the Chiang Dynasty, which would be rattled by a series of seismic events that set the stage for its demise in the 1980s. Loss of its UN membership and security council seat in 1971. US President Richard Nixon's visit to China and meeting with Mao in 1972. The death of Chiang Kai-shek in 1975. And the Big One in 1979: Washington abruptly ending its recognition of the ROC, nullifying the two old allies' mutual defence treaty and withdrawing its troops from Taiwan in favour of establishing diplomatic ties with China. Each shock further eroded the prestige of the House of Chiang, emboldening the activists inside and outside of Taiwan that sought its end.

The downturn for the Chiang dynasty began at the start of the decade when Chiang Ching-kuo, then vice premier, visited the US in 1970. Chiang dined at the White House at the invitation of Secretary of State William Rogers, who had met him the year before on a trip to Taiwan. The Chiangs were reportedly growing increasingly worried about the Nixon administration's unofficial overtures to Beijing – but not concerned enough to push for a meeting with the president.[1]

The mood in the US towards 'Free China' was changing. As Washington was warming up to Mao and the PRC, the bulk of the activities of the overseas Taiwan independence movement was shifting from Japan to the US. World United Formosans for Independence (WUFI) had just formed in New York, combining previously existing anti-KMT organizations in the US, Japan, Canada, Europe and Taiwan. WUFI organized protests that bedevilled Chiang throughout his visit, reaching a dramatic climax at the Plaza Hotel in Manhattan.

It was at this hotel, before a luncheon hosted by the Far East/America Council of Commerce and Industry, that a group of twenty-five or so protestors lay in wait for Chiang. As Chiang approached the hotel entrance, he was heckled by WUFI members. The confrontation escalated when a young man brandishing a black .25 calibre Beretta stepped towards sixty-year-old Chiang.

That man was Peter Huang, a thirty-two-year-old Taiwanese industrial engineering grad student at Cornell. Huang had been born in Hsinchu County during the Japanese era and had studied journalism in Taipei before his mandatory service in the ROC military. Luckily for Chiang Ching-kuo, Huang's marksmanship was impeded by NYPD Detective James Ziede, a member of Chiang's security detail, who grabbed the would-be assassin's arm a split-second before the gun fired. The bullet narrowly missed Chiang, instead piercing the hotel's revolving glass door.

As Huang was detained, he bellowed to the crowd outside the hotel entrance, 'Let me stand up like a Taiwanese!' This declaration would make him an instant hero to those who opposed KMT martial law and its colonization of Taiwan. Chiang emerged unhurt and, by all reports, unrattled by Huang's assassination attempt.

Photos of the two men at the Plaza Hotel shared the top of the front page of The New York Times the following day. In a dark suit with a blue tie and white pocket square, Chiang appeared to be forcing a smile that wouldn't quite form.[2] In a white T-shirt and denim jacket, Peter Huang held his head defiantly high, flanked by a pair of blue-uniformed NYPD officers holding his arms behind him. He and his brother-in-law and accomplice, the architect Cheng Tzu-tsai, would be convicted of attempted murder in New York, with both eventually skipping bail and fleeing to Europe.[3]

Huang had fired the first shot – literally and figuratively – of an increasingly fractious decade between the KMT and the people of Taiwan. (In 2012, in a display of how much the ROC had changed under democracy, Huang's undergrad alma mater, National Chengchi University – the former KMT party school founded in China – named him an 'outstanding alumnus'.)

By the beginning of the 1970s, it was also clear that Chiang Kai-shek's health was deteriorating. In 1972, he finally confirmed that Chiang Ching-kuo would be his successor, elevating him to premier. The Generalissimo would live for another three years, but would never be seen in public again. On 5 April 1975, a heart attack felled the eighty-seven-year-old, who had been debilitated by a long bout of pneumonia. Chiang Kai-shek had been the leader of the ROC for nearly half a century, and died with Soong Mei-ling and Chiang Ching-kuo – his closest confidantes and partners in power – at his bedside. To the very end, he never abandoned his delusional goal of retaking China, at least not publicly. Days before his death, Chiang penned a message to the citizens of the ROC. It was released by the government two hours after his passing.

'Just at the time when we are getting stronger, my colleagues and my countrymen, you should not forget our sorrow and our hope because of my death,' he wrote. Chiang urged the nation to 'recover the mainland and to restore our national culture'.[4]

Four days later, Chiang's replacement as president, Yen Chia-kan, placed a wreath atop his bronze coffin, which was loaded into a flower-adorned truck by his bodyguards. The coffin travelled through Taipei, from the Veterans' General Hospital north of the city to the Sun Yat-sen Memorial Hall downtown. As the truck and Chiang's coffin wended their way through the provisional capital of the ROC, hundreds of thousands of people lined the streets. A giant portrait of Chiang hung on the wall behind the altar. His coffin was placed on an inclined catafalque surrounded by flowers, candles, and the flags of the ROC and KMT. During the Christian funeral, the coffin lay open, showing Chiang in a Chinese robe, decorated with state honours. A children's choir sang 'Nearer, My God, to Thee'. Afterwards, Chiang was given a twenty-one-gun salute and a military band played Chopin's *Funeral March* before his coffin was transferred into a hearse. It would be taken to his mausoleum in the town of Cihu in Taoyuan County, where it was loaded into a black marble sarcophagus. Half a century later, the honour guard remains at the mausoleum.

Chiang's death might not have disrupted the running of the government, but it did mark a turning point for US–ROC relations. After the death of Franklin Roosevelt, Winston Churchill and Joseph Stalin, Chiang was the last to pass out of the Allied 'big four'. The respect that was accorded to him would not be simply transferred to his son.

The guest list for the funeral made immediately clear that both the ROC's – and Chiang's – prestige had dropped dramatically since the exodus to Taiwan. The US was the only major power that sent a representative to Chiang's funeral – Vice President Nelson Rockefeller. While his presence may have been a comfort to ROC officials, his words certainly were not.

Rockefeller said the late Chiang 'will long be remembered by the American people for his determination, courage and patriotism and for his great contribution to the Allied cause as comrade in arms during the Second World War'.

When it came to the new government under Chiang Ching-kuo, however, Rockefeller declined to show support. The most he could do was say, 'I know the people of the Republic of China have full confidence in those who have now been called upon to take up the burden of leadership.'

Regarding Washington's view of Taipei, Rockefeller made a statement that hinted at the shifting US–ROC relationship. 'Friendship will continue to characterize relations between us,' he said. ROC officials no doubt winced at such a tepid statement when it was known that Gerald Ford was planning a trip to China later that year. The fact that Rockefeller didn't even mention the mutual defence treaty between the US and ROC likely only added to concerns.

In Chiang's lifelong rivalry with Mao Zedong, he was the first to bow out, with his Communist bête noire giving up the ghost the following year in 1976. The vacuum left by these two formidable personalities would unleash powerful forces that would reshape the region, and the world.

Annette Lu and the Kaohsiung Incident

As with every year, the graduate law programme at Harvard in 1977 was brimming with bright, ambitious young minds. Among them were two promising scholars from Taiwan with very different backgrounds.

Jerome Cohen, who chaired the law school's graduate committee at the time, had first heard of Annette Lu from a star pupil of his who had studied in Taiwan three years earlier. Lu had already made a name for herself as a newspaper columnist and feminist activist – neither of which was the safest thing to be under Chiang-era martial law. While Lu was carefully critical of ROC gender politics, she had yet to go public with her support for Taiwanese independence – a stance which, if discovered, could have led to her imprisonment. Upon his student's advice, Cohen arranged a scholarship for Lu that would allow her to pursue a Master of Laws (LLM) degree at Harvard.

Another student from Taiwan had already earned his LLM at New York University and hoped to pursue his doctorate at Harvard. Cohen's junior colleague strongly recommended the student, which convinced the professor to admit a second promising talent: a young KMT cadre named Ma Ying-jeou.

Far from Taiwan, both Lu and Ma did little to hide their political ambitions back home or their opposing political views. Lu repeatedly approached Cohen about her suspicions that Ma was reporting her advocacy for Taiwanese independence. While Cohen was sympathetic to Lu's cause, he had to explain to her that her request that Ma be ejected from the programme wouldn't have any grounds in the US, where freedom of speech allowed both her and Ma to say whatever they liked as long as neither tried to impede the other's speech.

Ma served as Cohen's research assistant on human rights issues in China and Taiwan, impressing his professor with his progress in public international law. The academic year ended in late spring 1978, with an accomplished Ma moving on to the doctoral programme. Lu, however, was drawn back to Taiwan by the growing chatter that the Carter administration would switch Washington's recognition of the government of China from Taipei to Beijing.

For Lu, and the Taiwanese both in Taiwan and abroad seeking to overthrow the ROC, it was a big moment. De-recognition by the US would render Chiang Ching-kuo's government even weaker. Lu wanted to be involved in what would happen afterwards. She would end up getting her wish.

Lu and Cohen had a lengthy chat in his office in which they

discussed her options for helping the cause of Taiwanese self-determination. Cohen preferred that she work for independence outside of the KMT dictatorship by collaborating with the small but active international network advocating for Taiwan. The other option, which the impassioned Lu chose, was to go back to Taiwan to push for democracy within the system. This was the riskier option, but as events later would show, opposing Chiang Ching-kuo from abroad could also have deadly consequences.

Cohen's recommendation that Lu fight the good fight from abroad was no match for her desire to be back on the front lines of the democratic struggle in Taiwan. He feared that Lu would face imprisonment or worse from Chiang Ching-kuo's government. As she prepared to leave, Cohen told her, semi-jokingly: 'Don't worry – if they lock you up, we'll get you out.'[5] Cohen had been advising Senator Edward M. Kennedy, who had the ear of the human rights-focused President Carter, so such a statement would certainly carry weight. He would later come to regret his well-intentioned words of assurance.

Lu returned to Taiwan in the summer of 1978, just months before Carter announced he would drop a long-anticipated bombshell on Chiang Ching-kuo and officially recognize the PRC government in Beijing.

Under the KMT, Taiwan's political environment was tightly controlled, with the ROC party-state prohibiting the formation or operation of any opposition parties. However, Chiang Ching-kuo did allow contested elections for a handful of legislative yuan seats, including non-KMT candidates. Those candidates were called *Tangwai*, meaning 'outside of the party'. Initially used in reference to low-level elections as early as the 1950s, by the late 1970s, Tangwai had become synonymous with the Taiwanese self-determination movement.

The US decision to switch diplomatic recognition shook the political status quo in Taiwan. In response, a beleaguered Chiang Ching-kuo ordered elections postponed indefinitely and began cracking down on dissent and criticism of the government. The Tangwai movement saw its opportunity.

In August 1979, the first issue of *Formosa Magazine* was published by Tangwai leaders, quickly selling out its 25,000-issue print run. The following three issues would break Taiwan magazine sales records, no doubt heightening Chiang's anxiety – the Taiwanese independence movement finally had a platform and its message was resonating with a weary populace.

Getting back to her roots in journalism, Annette Lu served as *Formosa Magazine*'s editor. Her new job would be short-lived, as the magazine was banned in October 1979. *Formosa Magazine* branch offices around the island, effectively offices for Taiwan's democratic movement, were subject to harassment and attacks from police and gangsters – both of which were loyal to the KMT.

On 10 December 1979, in the southern port city of Kaohsiung, the movement-cum-magazine held a parade to commemorate International Human Rights Day. Turnout had been boosted by an incident the day before, in which two campaign trucks promoting human rights were stopped by police, who beat and arrested two volunteers. Police and soldiers were waiting for the protestors when they arrived. Aggressive behaviour by authorities combined with an angry crowd led to a violent clash between the two sides. This was followed by the raid of *Formosa Magazine*'s Kaohsiung office and the arrest of anyone remotely resembling a Tangwai leader. Tangwai members accused the government of planting agitators within the demonstration. Those arrested in what would become known as the Kaohsiung Incident were detained incommunicado for two months or longer, with widespread allegations of the abuse and torture of detainees.

Lin Yi-hsiung, the magazine's circulation manager and a major movement figure, was allowed a visit from his wife, Fang Su-min, in February 1980. It was apparent just by looking at him that he'd been beaten, after which Fang contacted the branch office of Amnesty International in Osaka, Japan.

The following day, Lin's mother and seven-year-old twin daughters were stabbed to death and his nine-year-old daughter was seriously injured. Although the police had Lin's residence under round-the-clock surveillance, they claimed to not know how the murders happened. The case has never been solved. To this day, it is widely believed to have been a chilling message sent by the KMT to intimidate independence-minded Taiwanese. If that is indeed what it was, it did not work.

A group of the eight most prominent magazine staff members, including Annette Lu and Lin Yi-hsiung, were put on trial at the military tribunal inside the Jingmei Detention Centre compound just south of Taipei. They were defended by a team that included the young lawyer and future president Chen Shui-bian. The military trial of the 'Kaohsiung Eight', as Lu and her cohorts had become known, was open to some members of the public and press.

At the time, Jerome Cohen was working as a corporate lawyer in Beijing and hoped to attend the trial as an observer. The trial date conflicted with business negotiations in Beijing that he would be involved with on behalf of his top client, Amoco. Cohen claimed that the scheduling conflict was not a coincidence, as the KMT had been eavesdropping on his phone calls from Beijing to Taipei. He sent a colleague to Taipei in his stead.

There was little doubt that the eight defendants would all be found guilty. The only question was how severe their punishment would be. The military tribunal meted out harsh sentences to all, including a twelve-year sentence for Annette Lu.

Cohen could do very little for his former student. As Lu languished in a military prison, her health deteriorating, Cohen saw few options for pressuring the KMT to release her. He wrote a critical piece for the *Asian Wall Street Journal* but knew it would have little impact.

It was only in late 1984, when a reckless KMT made a grave tactical error that jeopardized its ties with the US, that Cohen had the opening he needed to finally make good on his promise to Annette Lu.

The Murder of Henry Liu

It was the morning of 15 October 1984, and Bay Area resident Henry Liu was getting ready for work. Henry and his wife, Helen Liu (née Tsui) had moved to the US from Taiwan in the late 1960s and became naturalized citizens in 1973. The couple lived in Daly City on the southern edge of San Francisco and owned two gift shops in the area. One of their shops was in Fisherman's Wharf, a major tourist attraction, and the other was in San Mateo, to the city's south.

Liu was in his garage, the door raised while he loaded inventory into his car. Three Chinese men rode up Liu's driveway on bicycles, and drew guns. They fled after firing three shots: two into Liu's gut and one into his head. After hearing the shots, Helen Liu rushed to the garage, where she found her husband's lifeless body on the floor.

It quickly became apparent that Liu's murder was the politically motivated assassination of an American citizen on US soil by the ROC. The three assassins, it turned out, were members of the notorious Bamboo Union triad. They had been tasked with Liu's murder by none other than the ROC Ministry of National Defence's intelligence wing. US intelligence agents connected the gangsters with military intelligence in Taiwan by monitoring communications following Liu's assassination. As the gravity of killing a US citizen on American soil belatedly began to dawn on the KMT, the ROC government announced it had arrested three members of its military intelligence in connection with Liu's death.

But why would the KMT assassinate a humble shop owner in

the US? As it happened, Henry Liu was also a well-known author and journalist who was highly critical of the KMT and ROC government.

Liu was born in the coastal Chinese province of Jiangsu in 1932. When he was nine, his father was shot and killed by Mao's Communists. After fleeing to Taiwan in 1949, Liu was pressed into military service at the age of sixteen. After fulfilling his duties to the military, Liu transitioned to a career in journalism, initially working for ROC state radio before becoming a print reporter for the *Taiwan Daily News*, a major daily. He gradually built experience in international reporting while covering the Vietnam War before being sent to Washington to cover the US in 1967. Life in the US led to Liu growing increasingly disillusioned with the KMT. He decided to pursue American citizenship. Having secured citizenship for himself and Helen, Liu turned his focus to the KMT, writing critical and often sensational articles on the party under the pen name Chiang Nan for Chinese-language publications in San Francisco and British Hong Kong.

When he lived in Taiwan, Liu had met many of the people he would go on to write about. Helen told the February 1985 House Foreign Affairs Subcommittee investigating Liu's assassination that, during the 1950s, her husband had attended the academy for political officers – thought police – at Fu Hsing Kang College, which was run by Chiang Ching-kuo. Liu also met Chiang's younger brother General Chiang Wei-kuo, and Chiang Kai-shek's powerful right-hand man, General Wang Sheng. Wang oversaw the Political Work Department beginning in 1950, putting him at the top of a structure responsible for the countless arrests, disappearances and executions that characterized the earliest years of ROC martial law.

In his writing, Liu scrutinized all the most powerful members of the ROC aristocracy, including Chiang Kai-shek, Soong Mei-ling, Chiang Ching-kuo and Wang Sheng. According to Helen's testimony, these articles resulted in messages and visits from representatives of the ROC government during the 1970s and 1980s. In 1973, Wang Sheng personally sent a letter to Liu regarding a biography of Chiang Ching-kuo that the author was writing, telling him to 'move cautiously, and think twice' before publishing his book.[6] Henry Liu wrote Wang Sheng back from what he felt was the safety of the US. 'I'm living in America and I am independent,' Liu said. 'No one could tell me what I should write about.'[7]

At the time of his assassination, Liu was preparing to write a biography of K. C. Wu, who had served as mayor of Shanghai before the toppling of the ROC in China, and served as Governor of Taiwan post-exodus. Wu was also a rival of Chiang Ching-kuo. Wu resigned from his governorship in 1953 and left one month later for the US with his family, save one son, whom the Chiang regime had barred from leaving. Wu was expelled from the KMT and his son was eventually allowed to rejoin his family. The former KMT official would become a fierce critic of the Chiang regime's police state in Taiwan.

Before his murder, Liu had signed a contract with the Wu family giving him exclusive access to their extensive archives. In her statement before the House Subcommittee, Helen Liu said, 'Some people feel that it was official fear of this book that prompted Taiwan to order Henry murdered.'[8]

'Many people feel that Henry's planned full biography of Governor Wu would contain many more shocking and damaging revelations about President Chiang and his family,' she added. Helen blasted the Reagan administration for not condemning Chiang Ching-kuo's assassination of her husband, accusing it of seeking to avoid an inconvenient disruption of the Cold War alliance between Washington and Taipei.

While Reagan's White House remained quiet, Congress would not give the KMT a pass. New York Democrat Stephen Solarz, the chairman of the House Committee on Foreign Affairs Subcommittee on Asian and Pacific Affairs that presided over the hearing on Liu's murder, did not hold back:

> I cannot exaggerate the sense of outrage which the reported involvement of officials of the Taiwan Government in the murder of an American citizen on American soil provokes in me. [. . .] I know that there may be some disagreement among members of the committee and in the Congress over whether the United States should put pressure on repressive regimes abroad to respect the human rights of people in their own territory. But I am sure we all agree that the territory of the United States should not be allowed to become a hunting ground for foreign governments wishing to stifle dissent.
> [. . .]
> Part of my outrage stems from the knowledge that this is not the first time that Taiwan has abused the freedoms of individuals in the United States. In the past, there have been numerous credible charges of surveillance, intimidation, and harassment in the United States by agents of Taiwan's intelligence services, particularly with respect to Taiwanese students in our country.[9]

Students weren't the only targets of the increasingly unhinged ROC police state. Just three and a half years earlier, in 1981, the same House Subcommittee convened a hearing on the murder of Chen Wen-chen, an assistant professor of mathematics at Carnegie Mellon University.

Like Henry Liu, Chen had left Taiwan for the US after completing his compulsory military service. He had earlier received a bachelor's degree in mathematics at the prestigious National Taiwan University and would go on to get his master's and then PhD in statistics from the University of Michigan in Ann Arbor in 1978.

Chen's statistics professor at Michigan, Bruce Hill, told student newspaper *The Michigan Daily* that Chen was 'an outstanding student – the best that I'd seen in statistics in twenty-one years'.[10] Hill said he doubted Chen had time for political activities given how dedicated and busy he was as a student.

The Michigan Daily, however, disclosed in the same article that Chen was one of several Taiwanese students who had approached the newspaper five years earlier with a list of students that they claimed were secret agents of the KMT.

In private conversations, Chen was known to be highly critical of the KMT and supportive of a Taiwanese state. After the Kaohsiung Incident of 1979, he helped raise funds stateside for Annette Lu and her comrades in the Kaohsiung Eight.

Back in Taiwan, Chen's mother had managed to convince her son to return for a visit, despite his concerns about his safety, which would prove well founded. During his trip back to Taipei with his wife and one-year-old, the thirty-one-year-old Chen was detained by the feared Taiwan Garrison Command and interrogated for twelve hours. His body was later found at the foot of a tall building on the National Taiwan University campus. His spine was broken, as were thirteen ribs. One of his shoes reportedly had a folded bank note stuck inside, apparently to prevent Chen's soul from becoming a hungry ghost.

The Garrison Command initially stated that Chen had killed himself out of fear of imprisonment due to his seditious activities towards the ROC.[11] His family rejected the accusation, saying suicide was out of character for Chen, who had just started a family and had a promising academic career ahead of him in the US. After it became clear that nobody was buying Garrison Command's story, the government declared Chen's demise to be 'most likely accidental'.[12]

In September 1981, Cyril Wecht, an American forensic pathologist perhaps known best for his criticism of the Warren Commission's findings on the assassination of John F. Kennedy, flew to Taipei to examine Chen's corpse. Wecht's conclusion was markedly different from those offered by the KMT. In an article published by *The American Journal of Forensic Medicine and Pathology*, he declared that Chen had been murdered. He assessed that 'both the location of the body and the pattern of injuries are consonant with the explanation that Chen's body was held in a horizontal position over the railing of the fire escape and dropped onto the ground below'.[13]

In response, Taiwan's government did nothing – aside from revoking the press credentials of Taipei-based *Associated Press* reporter Tina Chou, a US national, after she filed an article citing Wecht's report.

The US House, however, did take action. It held a hearing on Chen's murder that concluded that the Chiang regime had killed him because KMT student spies had seen him attend certain events in the US. Two politicians, Stephen Solarz and Jim Leach, who would later be on the Subcommittee looking into Henry Liu's assassination, seemed personally insulted by the KMT's brazen actions.

'The Taiwan government has been effectively on notice since 1981 that any act taken by their agents against any individual in this country who is engaged in the lawful exercise of his or her civil and constitutional rights would seriously jeopardize the warm relations between our governments,' Leach said during the Henry Liu hearings.

'How the Taiwanese authorities respond to US concerns with regard to the Henry Liu case may well decide the future course of US–Taiwan relations,' he added. 'The conduct alleged in the case before us, of government-sanctioned murder, is not the conduct of friends.

'As long as the broad brush of national security can be used to gloss over the excesses of the state, there can be no guarantee in the future that murderous acts will not reoccur,' Leach continued. 'The murder of Henry Liu must be seen in this larger context, and the Government of Taiwan urged once again to repeal martial law.'[14]

California Democrat Norman Mineta was equally blunt:

'I say to my friends in Taiwan that US law does not allow us to sell arms to a country when there is a systematic pattern of intimidation or harassment against US citizens,' Mineta said. 'These are supposedly our friends. We sold them $760 million in arms in 1985,' he continued. 'I think the time has come to tell the so-called friends of ours to take their intelligence operatives and recall them home.'[15]

The State Department was also losing its patience with Chiang Ching-kuo's increasingly rogue government's apparent disregard for American sovereignty. William Brown, the Deputy Assistant Secretary for East Asian and Pacific Affairs, also spoke before the Subcommittee and was asked by California Democrat Tom Lantos when the State Department learned of Henry Liu's murder:

BROWN: *We received a full briefing, what I would characterize as a comprehensive briefing in December, sir.*

LANTOS: *What was the reaction within the department upon receiving that?*

BROWN: *Outrage.*

LANTOS: *How fully was this outrage communicated to the Taiwan authorities?*

BROWN: *Immediately and vigorously.*

LANTOS: *And at what level?*

BROWN: *Those communications ran through the auspices of the American Institute in Taiwan, with its counterpart at the highest level.*

LANTOS: *Is it reasonable for this committee to conclude that the Secretary of State was fully briefed on this event?*

BROWN: *It is, sir.*[16]

Annette Lu's Release and the End of the Chiang Dynasty

Shortly after the assassination of Henry Liu in October 1984, when US–ROC relations were at their lowest level since the Second World War, Jerome Cohen boarded a plane at Hong Kong's Kai Tak Airport and flew to Taiwan. In order to help his former student Annette Lu, Cohen sought the assistance of another former student, Lu's former nemesis at Harvard, Ma Ying-jeou.

After returning to Taipei with his Doctorate of Juridical Science from Harvard, Ma became a rising figure in the KMT, serving as a deputy secretary-general for the party, as well as President Chiang Ching-kuo's English interpreter. Cohen believed that Ma would understand the importance of repairing relations with Washington, which had become progressively worse following Henry Liu's assassination, as well as the earlier trial of the Kaohsiung Eight and the murder of Chen Wen-chen.

When Cohen met with Ma, he made the case that, given the poor state of bilateral ties, releasing Annette Lu was low-hanging fruit for the ROC that would temper Washington's displeasure. After listening politely to his former professor, Ma asked him to return to Taipei in a fortnight.

Two weeks later, Cohen made the short flight back. He was received by Ma, who informed him they would meet with Lu at the prison hospital where she was receiving cancer treatment.

'That meeting was probably the most emotionally moving one I had ever had,' Cohen writes in his memoir, *Eastward, Westward*. 'Ma looked healthy and handsome in a dark three-piece suit, and his mind was razor sharp. Annette appeared more forlorn than I had ever seen her, far from her former vibrant self and rather lethargic in a dull prison dress.'[17]

Cohen apologized for not following through on his promise to help Lu earlier. She had no recollection of his pledge, but she did acknowledge that she had not anticipated the level of trouble in which she would find herself. The three chatted for a little over an hour before the two men left Lu and made their way back into downtown Taipei.

A few days later, Ma checked in with good news: Annette Lu would be released from prison. There was just one caveat. Lu would not be allowed to stay in Taiwan for her first year of freedom.

Cohen arranged for Lu to return to Harvard for another year, this time as a fellow at its Center for International Affairs. Lu would be the third persecuted East Asian democratic leader Cohen would help get into the centre – the previous two being Senator Benigno Aquino of the Philippines and Kim Dae-jung of South Korea. Aquino was later assassinated in 1983 at Manila International Airport upon returning from Boston. Aquino's wife, Corazon, and his son, Benigno (aka Noynoy), would both eventually be elected Philippine president. Kim Dae-jung would later win the Nobel Peace Prize and be elected South Korea's president.

After Lu's release in 1985, Cohen remained invested in US–Taiwan affairs, especially in the case of Henry Liu. The ROC government did not have an extradition treaty with the US – which is still the case today – so Washington had little leverage. There would be two trials in Taiwan: one for the Bamboo Union gangsters, and another for Admiral Wang Hsi-ling – who had ordered the hit on Henry Liu – and his two closest aides.

But there would be a third trial as well, this one in the US. Helen Liu had retained the services of a lawyer named Jerome Garchik, who was another former student of Cohen's. Garchik called his former professor and told him Liu was suing the ROC government for the wrongful death of her husband. Cohen's firm – Paul, Weiss – was highly respected in civil litigations, and Garchik asked if Cohen would be able to deal with things in Taiwan.

Cohen accepted and Paul, Weiss took on the case on a pro bono basis. The firm's lawyers argued successfully that the ROC could be held responsible in a US court despite its sovereign status and lack of formal relations with the US. After that, the burden to uncover evidence implicating Taipei in Liu's murder fell primarily on Cohen's shoulders.

The ROC government refused to extradite any of the suspects from Taiwan to the US, but that didn't prevent it from coming under massive political pressure from its most important security partner. International and domestic backlash, combined with growing discontent within Taiwan towards the Chiang government's authoritarian tactics, pressured the ROC to proceed with trials for the suspects within its jurisdiction.

Cohen sought to intertwine Helen Liu's lawsuit in California with the criminal proceedings in Taiwan against the Bamboo Union hitmen accused of executing the assassination. Cohen would need help from a Taiwanese attorney, but none of his former students from Taiwan wanted to touch the case given its highly sensitive nature. He was eventually introduced to Frank Hsieh, a Taipei city councillor.

Hsieh had studied law at National Taiwan University and Kyoto University, was bright, ambitious and renowned for his readiness to confront the KMT. In Cohen's words, 'Frank seemed to be a very competent and energetic litigator, and we proved to be an effective team.'

Cohen was initially barred from attending or taking part in the trial of the Bamboo Union assassins. He received help from the court of Taiwanese public opinion, however, when local media criticized the judge for blocking the bereaved Helen Liu's representative from the courtroom proceedings. Cohen was then allowed to observe the trial, which was held under maximum security. After each day's hearing, he would speak to television and print reporters, slamming the trial judge's unwillingness to allow him to participate in the case itself.

As expected, the assassins were convicted and subsequently

appealed the ruling. During this time, Cohen and Hsieh also appealed the decision to bar Cohen from the trial. They won, and the trial would be held all over again, but now the two lawyers were allowed to question witnesses and defendants. They could now extract important information and testimony that would be vital in Helen Liu's suit in California. Hsieh obtained a crucial police interrogation transcript that would help confirm the government's role in arranging the murder of Henry Liu. The Bamboo Union hitmen would be convicted once more.

Admiral Wang and his two aides were not prosecuted in criminal court, but rather in military court. Cohen was allowed to attend the case's public hearing, in which Admiral Wang denied guilt for the murder. His unorthodox defence was that the apprehension of the murderers proved that there was no way the Defence Intelligence Bureau could be involved, as the bureau was much too competent to be caught. Speaking before judges and a courtroom with a hundred or so observers, the Admiral spoke with a chilling arrogance: 'You know, your honours, we are professionals,' he said. 'Ten or twenty years ago, we used to do things like this all the time, and were never discovered.' A shocked silence engulfed the room.[18]

As with the Bamboo Union boss Chen Chi-li, Admiral Wang was convicted and sentenced to life in prison. Both men were subjected to a VIP-style incarceration that was closer to house arrest and were released a few years later. Regardless, Helen Liu's litigation against the ROC in an American court was successful. Realizing just how much damage the murder of Henry Liu had caused its reputation – almost certainly more than any book Henry Liu could have written would have done – the KMT settled for US$1 million, a large sum for such a case in the mid-1980s. While a million dollars was no big deal for the KMT, which in 1986 was famous for being the world's wealthiest political party, the damage done to KMT rule in Taiwan was much more substantial.

Kathrin Hille of the *Financial Times*, one of the sharpest foreign reporters on Taiwan, would later write that the government-ordered killing of Liu on American soil was 'the most prominent example of the KMT's cooperation with gangsters in upholding its dictatorship'.[19] Most prominent, perhaps, because it was witnessed by the international community as well as the Taiwanese people. It spurred a more critical examination of the KMT's rule and its implications for human rights, democratic reforms and the rule of law both at home and abroad.

Indeed, in 1986, the foundations of Taiwan's democratic future were being laid. Frank Hsieh and his colleagues within the Tangwai established the Democratic Progressive Party on 28 September. It was a bold move towards formal opposition under a regime that banned the formation of any opposition parties.

All of this set the stage for what must have seemed unimaginable for many Taiwanese. On 7 October, less than two weeks after the DPP's founding, an ailing seventy-six-year-old Chiang Ching-kuo met with Katharine Graham, chairwoman of the board of *The Washington Post*, as well as two editors from *Newsweek* and the *Post*. With Ma Ying-jeou seated next to him and serving as interpreter, Chiang told the American media representatives that he intended to lift martial law, which had been in place since 1949. The government would end the trials of civilians in military courts while lifting restrictions on personal freedoms, he said without elaborating.

On 15 July 1987, Chiang followed through on his promise, ending martial law in Taiwan after thirty-eight years. Half a year later, on 13 January 1988, he would die of a heart attack, ending the Chiang dynasty in Taiwan. He would be laid to rest in a mausoleum of his own, near his father's in Cihu. And with that, more than four decades of Chiang rule in Taiwan had ended.

The end of authoritarian rule in Taiwan would open the way for political reform and eventual democratization. The KMT would be pitted against the upstart DPP, shaping the island's political landscape and determining its future leaders. Many of those on both sides of the fight in the 1970s and 1980s would later resurface as major electoral candidates. In 2000 and 2004, Annette Lu would be elected vice president alongside Chen Shui-bian, one of her former lawyers in the Kaohsiung Eight trial. In Taiwan's 2008 election, Ma Ying-jeou would defeat Frank Hsieh to become president.

CHAPTER 7

The Three Immortals

Between 2019 and 2022, the Taiwanese independence movement lost three of its most senior and influential figures. Ranging in age from ninety-seven to one hundred years old, their lives spanned the seismic geopolitical shifts experienced by the Taiwanese people over the past century. From living through the Japanese colonial period and KMT martial law to the recent democratic present, the lives of Lee Teng-hui, Su Beng and Peng Ming-min also reflect the resilience and perseverance of the Taiwanese people in their pursuit of self-determination.

In their own way, each of these individuals helped awaken a new democratic Taiwanese consciousness. They are three crucial figures in the evolution of the Tangwai movement into the DPP, which is now Taiwan's majority party.

The death of Chiang Ching-kuo in 1988 marked a critical turning point for all three men. For Su Beng and Peng Ming-min, it signified an opportunity for them to return to Taiwan from their exile abroad. For Chiang's vice president, Lee Teng-hui, it propelled him into becoming the first Taiwan-born president of the ROC. His presidency would begin in a post-martial law era, in an environment where little had changed on the ground. Twelve years later, he would finish his one term as the first elected president of a democratic Taiwan.

Mr Democracy: Lee Teng-hui

15 January 1923 – 30 July 2020

With the death of Chiang Ching-kuo just two weeks into 1988, an isolated Taiwan was thrust into an era of post-Chiang uncertainty. Nobody inside or outside Taiwan knew what to expect.

Chiang's obituary in *The New York Times* noted that he had been succeeded by the Taiwan-born Lee Teng-hui, who had worked his way up to the apex of the KMT hierarchy. 'Mr. Lee is expected by some analysts to serve out Mr. Chiang's term, which ends in 1990,' the obit read. 'He is well liked on the island, but is regarded as possessing less strength of will, leadership ability and charisma than did Chiang Ching-kuo.'[1]

As it turned out, Lee possessed *more* strength of will, leadership ability and charisma than his predecessor. He would prove to be one of the most impactful Asian leaders of the late twentieth century.

Lee inherited a complex political system in which his initial years as president saw him elected not by a public vote, but by members of the National Assembly. The assembly, representing each province of China, had not seen fresh elections since 1947 due to the ROC's loss of control over China. The men who had been elected four decades earlier were ordered by the Judicial Yuan to continue to hold office until new elections could be held in China. So, when Lee completed the late Chiang Ching-kuo's term in 1990, he was elected by a group of elderly men who had been born in China in the first half of the century. Some arrived to vote in wheelchairs. One famously entered carrying his urinary catheter bag.

Lee was 're-elected', unopposed, by a vote of 641 to 0.

This was red meat for Taiwan's democracy movement. The DPP had illegally nominated their own presidential candidate, Huang Hua, and during the election students from National Taiwan University, Lee's alma mater, occupied the vast square in front of Chiang Kai-shek Memorial Hall to demand constitutional reform, the end of the National Assembly and direct elections. They would be joined by students from twenty other universities, in what would become known as the Wild Lily Movement. A core of an estimated 6,000 students occupied the square, with an additional 10,000 or so fellow citizens encircling them for protection. Some protestors went on hunger strike.

This was only ten months after the PRC leader Deng Xiaoping had ordered the bloody clearing of Tiananmen Square and other protest sites across China in June 1989. In Taipei the possibility of government violence against the protestors was very real – four decades of martial law had ended less than three years earlier, and Taiwan was not the safe space for political expression that it is today. But rather than clear the square like Deng, Lee decided to meet with a group of fifty-three student representatives immediately after he was sworn in for his first full term as ROC president.

Fan Yun, a twenty-one-year-old sociology student who had drawn media attention for her impassioned speeches and media interviews, was part of the delegation that called upon Lee. They met at the presidential office, just around the corner from the square the students were occupying. Like Fan, Lee had been a student leader at National Taiwan University and had organized a 1947 protest against the US following the rape of a Peking University student, Shen Cong, by two American soldiers in Beijing.

Fan and the other students didn't know whether Lee would be friendly to their cause. As it turned out, he was.

'President Lee was very welcoming – he wanted to convince us that he had the same values and sought reform,' Fan, now a DPP legislator, told me when Lee was still alive. 'He is a very remarkable person. He accelerated our democratization.'

Lee agreed to hold a national conference later that year to discuss how to proceed with democratization. Two years later, in 1992, Taiwan would hold its first direct legislative elections. Six years later, in 1996, they would vote into office their first directly elected president: Lee Teng-hui. Lee and his running mate, Lien Chan, cruised to a comfortable victory with 54 per cent of the vote, compared to DPP candidates Peng Ming-min's and Frank Hsieh's 21 per cent. Two former KMT members running as independent candidates, Lin Yang-kang and Chen Li-an, received the remainder of the votes.

The CCP didn't know Lee well, given that he was born and raised in Taiwan – he was the first ROC president who wasn't born in China. But Beijing saw enough of pre-democracy Lee to know that it didn't trust him. When he ran for president in 1996, China fired missiles into Taiwanese waters until US President Bill Clinton dispatched two carrier groups from the Seventh Fleet to the strait.

Over the course of his twelve years in office, Lee oversaw significant political and social changes in Taiwan. Among them were the open resurgence of Taiwanese identity, the abandonment of the aspiration to 'retake the mainland', and the birth of a multi-party democracy. During the Lee era, Taiwan also saw the implementation of universal healthcare and the first public discussion of a topic that had been repressed for decades: the 228 massacres.

In 1990, he ordered the establishment of a 228 Incident Task Force and permitted 228 to finally be mentioned in high school textbooks. In 1995, in a historic act of acknowledgement, Lee addressed 228 publicly, offering a formal apology on behalf of the government for its past atrocities. Speaking to a solemn audience at the dedication of the 228 Memorial near the old radio station from which the call for rebellion had been transmitted, Lee sought to acknowledge the massacre in the hope that it would help Taiwanese people of all backgrounds move forward together.

'Today, the family of the victims will listen with their own ears as I, as a public servant of the country, accept the responsibilities of the government's past mistakes and offer my deepest apologies,' Lee said.[2]

It was a remarkable acknowledgement that would open up public discourse on the once verboten topic. As discussed earlier, there were a paltry four mentions of 228 in Taiwan's biggest newspapers before 1984, all of which were pro-government in tone and published in the *Taiwan Shin Sheng Daily News* before 1958. It wasn't until 1987 – four decades later – that the first article critical of the government's slaughter of Taiwanese would appear in print, running in the *United Daily News*.

Lee also made significant contributions towards the establishment of a new Taiwanese identity and enhancing Taiwan's international relations. One of his significant achievements was strengthening the now-unofficial bonds between Taiwan and the US.

As ROC president, Lee also boosted ties between his government and Japan. As a colonial subject, Lee was fluent in Japanese and had served in the Imperial Japanese Army from 1944 to 1945 (and, twice in the following two years, would briefly join the CCP as well).[3] Lee's efforts also extended to Europe, where he increased exchanges with Germany, whose own experience in dealing with the ghosts of the past would prove helpful to Taiwan's attempts to do the same.

Despite China's threats and military manoeuvres in 1995 and 1996 that are known as the Third Taiwan Strait Crisis – the first two occurred in the 1950s, and were centred around islands near China's shores that the ROC had retained control of during its retreat to Taiwan – Lee also further opened the door to China. (Family visits to China were initially permitted in the early 1980s and eventually permitted outright in 1987.) His China policy led to countless manufacturing and other low-cost labour jobs moving across the strait from Taiwan, where labour costs were still low and the economy was only just beginning to morph into the behemoth that it is today. It also became easier for Taiwanese people to go to China. Many Taiwanese *waishengren* were able to visit family they had not seen since the 1940s. Ironically, opening up cross-strait travel boosted the sense of a distinct Taiwanese identity – *benshengren* and *waishengren* alike had been told for decades that they were a part of China but, when they went there, the country that had spent almost half a century under Communist rule seemed like an alien world.

As a *benshengren* of Hakka descent who came of age under Japanese rule followed by KMT martial law, Lee would champion the New Taiwanese identity. For Lee, this concept was informed by the German notion of *Gemeinschaft*, or individuals bound by common norms, often due to sharing the same space and beliefs. One of his biggest moves as president was getting rid of the Taiwanese provincial government, effectively bringing the concepts of the ROC and Taiwan closer together and undercutting the Chiangs' claims that Taiwan was but a province of the ROC.

The deeper into his presidency that Lee got, the more Taiwanese and less ROC he appeared, even leading energized chants of 'Long live Taiwan! Long live democracy!' at packed rallies. The culmination of Lee's efforts in championing democracy came with the transition of power in 2000.

James Soong, a charismatic figure within the KMT and a staunch believer in the ROC, was widely regarded as the party's best chance to retain the presidency. His popularity and commitment to ROC principles had made him a favoured contender among the party faithful. In 1988, Soong had personally appealed to the Palace faction of the KMT, which deeply distrusted native-born Taiwanese, to approve then-Vice President Lee's succession of the late president Chiang Ching-kuo. But in 1999, to the shock of many, Lee opted to nominate Vice President Lien Chan for the presidency, rather than Soong. This rebuff by the president he had helped get into office a decade earlier was too much for Soong, who instead ran as an independent, effectively splitting the KMT 'blue' vote.

The former Taipei mayor and lawyer for the Kaohsiung Eight, Chen Shui-bian, threaded the needle, winning for the DPP with his running mate Annette Lu. With over 82 per cent voter turnout nationwide, Chen and Lu won with 39.3 per cent to Soong's 36.8 per cent. A humiliated Lien received only 23.1 per cent. History had been made. The pro-Taiwan 'green' camp led by the DPP had democratically wrested the ROC presidency from the KMT for the first time. KMT hardliners would never forgive Lee for splitting the vote.

After moving to Taiwan in 2015, I began trying to secure an interview with Lee Teng-hui through his eponymous foundation – gently, of course, as Lee was in his nineties and there was no public indication of his physical or mental condition. Every few months, I would call or email to remind his foundation's staff that I was keen to speak with him, if possible, at a time and place of his convenience. After three years, in May 2018, the efforts paid off – Lee was willing to talk with me at his home. It would turn out to be his last media interview.

Lee's modest but well-appointed flat was nestled in a discreet gated community near the National Palace Museum and its imperial treasures. Upon my arrival, the door opened and a Taiwanese man and woman in their thirties emerged, each wearing white, untucked collared shirts. Both polite and friendly, they were staff tasked with caring for the ninety-six-year-old former president. After a smiling exchange of names and business cards, it quickly became apparent that there was an expectation gap between the two sides.

I had arrived with a photographer who had flown over from Hong Kong. Lee's staff sat us down at a small folding table in an empty spot in the garage and provided us with a pot of tea. They told us that Lee was not dressed appropriately for a photo – could we maybe not take one, they asked. We explained that we were happy to try to find a fix. They suggested supplying a photo of him for us later, while we suggested maybe he could just put on a dress shirt and we'd shoot him from the waist up. The two wanted to help us but also didn't want to inconvenience Lee (nor did we). An uneasiness descended upon our amicable discussion of options, and they went upstairs to see what they could do.

We waited for about half an hour, growing slightly more nervous as the minutes passed. They had told us that it wasn't easy for Lee to change clothes or receive guests. We wondered: would he be able to handle an interview? Were we being rude by pushing for a photo? I had originally received permission to be joined by a photographer from the foundation staff, but perhaps the people running Lee's day-to-day activities didn't get the memo.

After we'd finished off the contents of the teapot, the staff returned to the garage with good news: Lee Teng-hui would see us now.

We entered and took a small staircase up to the main floor, which revealed a contemporary home with dark wooden floors and minimalist Japanese decor. In the room across the main chamber we had just entered, I could see a man sitting in a reading chair.

'Mr Horton!' Lee's voice boomed in relaxed, natural English, as he extended his hand towards two chairs across from him. 'Please, have a seat.'

His thick white hair was trimmed short. He wore a grey tweed sport coat with a white pocket square over a white shirt buttoned to the top with no tie. He was smiling and seemed relaxed.

The man who greeted us downstairs sat beside Lee to interpret. My Mandarin was sufficient for interviews but Lee was most comfortable speaking in Taiwanese. He would pepper his answers with occasional terms or phrases in Japanese and English. Speaking slowly and deliberately, we started by discussing the 1990 Wild Lily Movement, and his promise to the student protestors that set the stage for Taiwan's democratization.

'I've always cared about our students,' Lee said, reminding me of his time as a leader of student protests against the US seven decades earlier.

I asked him what he was most proud of during his time in office, assuming it would be delivering the democracy for which generations of Taiwanese had fought and died. His answer was quick, and somewhat unexpected: 'Nationalizing the military.'

It made sense. The ROC military had previously sworn its loyalty to the KMT, just as the PLA pledges allegiance to the CCP today, embodying Mao's maxim: 'Political power grows out of the barrel of a gun.'

'It was nationalization of the military that was the true starting point of all of my democratic reforms,' Lee told me. 'National defence is the crucial foundation for national security, and it is the most important symbol of national sovereignty. My democratic reforms were necessary to make a sustained push for the military to never again belong to an individual or political party.'

As mentioned earlier, Lee had responded to the Wild Lily Movement's demands by organizing the National Affairs Conference of June 1990, and it was this conference that resulted in the effective nationalization of the military, as well as the dissolution of the farcical National Assembly and the establishment of direct elections for all public officials.

As Lee told me, 'We achieved a bloodless, quiet revolution.'

Over the course of an hour, Lee voiced his concerns about local governance issues undermining popular support for democracy, the Trojan horse of relying on China as a trade partner and the growing threat of the PLA to Taiwan and the wider region.

When I asked him whether he supported Taiwanese independence, as both the KMT and CCP have alleged, Lee gave an answer that epitomizes the tightrope that Taiwan must walk: one that requires pragmatism and acknowledgement of unpleasant geopolitical realities, but not at the cost of dignity.

'The Republic of China in Taiwan and the People's Republic of China both exist individually – the only important point is the ability to exist under the name of Taiwan,' he told me. 'I've actually never advocated Taiwan independence, because Taiwan is essentially already independent,' he added. 'There's no need to make any announcements that will spark disputes in the international community.'

Lee's cross-strait approach lives on today. Both current President Lai Ching-te and his predecessor, Tsai Ing-wen, have made clear that they do not seek 'independence', as Taiwan is already an independent country. But if this is true, when did Taiwan become independent? Former Chief Justice Hsu Tzong-li has argued that an amendment included in the post-martial law constitutional reforms of 1991 created the current reality of 'state-to-state' relations across the strait.[4] Based upon this argument, it was during Lee's time as president that Taiwan not only became democratic – but a sovereign, independent state.

The Marxist: Su Beng

9 November 1918 – 20 September 2019

As with Lee Teng-hui, I was fortunate enough to sit down with the writer, restaurateur and revolutionary Su Beng shortly before his passing.

When I met him in New Taipei City in March 2019, Su was a century old and had outlived his entire family. He had a fine mane of white hair, a wispy goatee and was wearing a light hunting jacket with corduroy collar and cuffs, his blue jeans hiked up high on his waist. He was rather bundled up for being indoors, but not just because of his age – most homes in Taiwan don't have heating and the humid cold of the short winters in the country's north permeates indoor spaces.

After we shook hands, he gave me a business card from the Office of the President that indicated he was an adviser to the then-president, Tsai Ing-wen.

'I'm too old,' he lamented. His left eye was cloudy and mostly closed, but his right eye had a spark in it as he sized me up. 'It isn't easy for me to hear or talk or get around these days.'

Like most *benshengren* of his age, Su preferred to speak Taiwanese. He occasionally switched to Mandarin, while also sprinkling in some Japanese and English. He was living at the home of a young couple who had just had their first child. The husband, Lan Po-shih, helped interpret for me. I would ask my questions in Mandarin and Lan would lean towards Su, shouting the Taiwanese equivalent into his ear.

During our interview, Su had occasional coughing fits that shook his whole body. I told him several times that we could stop at any time. *Buhaoyisi*, I apologized. He waved away my apologies in a manner that suggested both that he'd been through worse before, and that he wanted to speak because he knew his time was near. Indeed, he would pass away just six months later.

Born under Japanese rule with the Taiwanese name Lîm Tiâu-hui to a wealthy family in Taipei in 1918, Su was educated in Chinese culture and Confucianism by his mother. As he grew up, he was influenced by the lively discussions his father had with friends who opposed Japan's colonization of Taiwan.

Su would excel in the Japanese education system and would go on to study economics and politics in Tokyo at the elite Waseda University in the late 1930s. He was introduced to Marxism and the CCP through a reading group organized by Chinese classmates.

There, he began to envision liberating Taiwan from Japanese rule. For Taiwan to be free, he concluded, China must first be freed from Japan's imperialist occupation.

Su travelled to northern China, where he would spend more than seven years in service of Mao's revolution, sometimes engaging in guerrilla warfare with the Red Army against the Japanese, other times spying on the KMT – which was still years away from seizing Taiwan in 1945.

Before Mao declared the founding of the PRC in 1949, the Communists had been conducting a bloody land reform campaign that encouraged villagers to speak out against landlords and other oppressors – real or imagined. Victims of the mass movement were publicly humiliated and executed, with estimates of those killed ranging from the hundreds of thousands to the millions. Su became increasingly disillusioned with the party.

'The Communist-liberated area back then lacked an international perspective,' he told me. 'Although [future PRC leaders] Zhou Enlai and Deng Xiaoping had studied in France, most of the cadres came from a peasant background. At that time, I remember thinking that it would be quite difficult for the CCP to occupy Taiwan.'

By 1949, the success of the Communist revolution was all but certain and the party was pressing Su to join. Having witnessed countless executions of Chinese by Communist forces, he realized the party's true ideology was not Marxism but control through fear. As Mao said in his most famous quote used for justifying political violence: 'A revolution is not a dinner party.'

'Why should you need to kill so many people to move things forward?' Su asked.

Instead of joining the party, he hopped on a boat heading from the northern Chinese port city of Qingdao to Taiwan, which by then had been occupied by the KMT for four years. Things in his homeland, he would learn, were not going well.

Upon Su's return to Taiwan, he was profoundly affected by the KMT's oppressive rule. His deep-seated belief in the necessity of an independent Taiwanese state drove him, alongside some like-minded associates, to conspire to assassinate Chiang Kai-shek. They began amassing a weapons cache. Their bold plan, however, was discovered by the KMT in 1952, forcing Su to flee. He went to the nearest major port, Keelung, where he boarded a boat carrying a cargo of bananas to Japan. The Japanese Empire was gone and Su was no longer a Japanese subject, so he was arrested upon his arrival for violating immigration rules.

The KMT demanded Su's repatriation, undoubtedly planning to execute him. The Japanese government allowed him to stay. This act of clemency opened a new chapter for Su Beng in Tokyo, where he reconnected with Hiraga Kyoko, a former girlfriend from his time in China during the war. In 1953, the couple opened New Gourmet, a Taiwanese restaurant in Tokyo's Ikebukuro neighbourhood. He told me that Taiwan-style *dalumian* – wheat noodles served in a thick gravy – and boiled dumplings were among its top sellers.

Japan was recovering from the ravages of war and the economy was picking up. As disposable incomes rose, the restaurant emerged as a popular neighbourhood eatery. Su would go on to run it for four decades, with the steady cash flow it generated helping to underwrite a larger endeavour: the overthrow of the KMT.

His project had both theoretical and practical components. First, the Taiwanese masses needed awakening. When he wasn't working at the restaurant, Su was poring over Japan's imperial archives, which were filled with documents about Taiwan. He began researching and writing a three-volume book on Taiwan's history, the first of its kind to offer a non-colonial vantage point. Through this project, he hoped to enlighten the Taiwanese people about their own history, adopting the nom de plume Su Beng – the Taiwanese pronunciation of the characters for 'history' (史) and 'clarity' (明), pronounced 'Shih Ming' in Mandarin. His choice of moniker proclaimed his goal of making Taiwanese see history clearly.

In 1962, Su's book, *Taiwan's 400-Year History*, was first published in Japanese, a language that most Taiwanese could still read. However, it faced censorship in Taiwan under Chiang Kai-shek's regime. Eighteen years later, in 1980, it would be published in Mandarin by a California-based publisher, Paradise Culture Associates. This publication coincided with a critical moment in Taiwanese history following the severance of US–ROC diplomatic ties and the trial of the Kaohsiung Eight.

The crux of the book's argument was that a distinct Taiwanese identity had been forged by four centuries of serial colonization. It resonated deeply with many Taiwanese and made Su a leading figure in the movement against the KMT party-state. The book's revelations were particularly poignant for Taiwanese students in the US, many of whom were encountering the details of the 228 Incident for the first time. The US, with its large Taiwanese diaspora, became a fertile ground for Su's activism, allowing him to engage with and rally support from students and wealthy expatriates alike. Funds raised would finance the second part of his project: the revolution.

In 1967, Su founded the Taiwanese Independence Association, which sought to destroy the ROC apparatus. Taiwanese customers would stop by his restaurant in Tokyo and, after enjoying a meal, would adjourn upstairs, where Su taught them about arson, bomb-making and guerrilla warfare. He claimed that the number of people he trained was in the thousands. The group took credit for attacks against symbols of KMT authority, such as police stations and military trains.

The death of Chiang Kai-shek in 1975 led Su to adopt a new, nonviolent approach, deciding that realism was more important for his cause than idealism. In 1992, the move towards democracy with the first direct elections for the Legislative Yuan convinced him it was time to go home, again.

He would return to Taiwan in 1993. This was a Taiwan that was six years out of martial law and exploring previously forbidden questions about its identity and future. Despite being seventy-five years old, Su was relentless: speaking at rallies, lending support to progressive social movements and funding a small fleet of trucks that drove around major cities with loudspeakers, calling on listeners to peacefully rise up and establish a Taiwanese state.

In the last stages of Su Beng's life, the political landscape of Taiwan was markedly different from the colonial-era Taiwan of his youth. When I spoke with him, the DPP controlled both the presidency and the legislature. Although the ROC still existed, it was largely in Taiwanese hands. A quarter to a third of society supported the KMT and the ROC. Xi Jinping was taking an increasingly hard line towards Taiwan and unification, but Su remained optimistic.

'Taiwanese society has its issues – it's not united,' he told me. 'But when threatened it will come together.'

That said, he was concerned about the meteoric rise of Han Kuo-yu, the KMT mayor of Kaohsiung who was a fierce proponent of closer ties to China. Han looked poised to potentially defeat the DPP's Tsai Ing-wen's 2020 re-election bid, which was ten months away. Just before Su died of pneumonia in September 2019, the election, only four months away, weighed on his mind.

'Taiwan is bound to prevail,' Su reportedly said on his deathbed, adding one last endorsement: 'Tsai Ing-wen must win.' She did.

The Lobbyist: Peng Ming-min

15 August 1923 – 8 April 2022

In the pivotal moment before Taiwan held its first direct presidential election in 1996, the country witnessed its first live televised presidential debate. In what must have startled many viewers, the third of four candidates, Peng Ming-min of the DPP, addressed the television audience in Taiwanese.

Wearing a light navy suit with a white polka-dotted tie, the bespectacled seventy-five-year-old got down to business, criticizing the three networks that were broadcasting his speech as being mere extensions of the KMT propaganda machine. He then drew parallels between Taiwan's 228 Memorial and the Vietnam Memorial in Washington, DC, to underscore that Taiwan's memorial lacks any sort of historical explanation. The ROC government under the KMT, he said, has fostered a lack of historical awareness among the Taiwanese people.[5]

Then he pivoted to the KMT itself, which at that point had ruled Taiwan for fifty-one years, thirty-eight of which had been under martial law:

> *Under the arrogant rule of KMT officials, is there any way that they have the genuine love for this land that the average Taiwanese person has? Do they know the reason why the ancestors of the Taiwanese people crossed such treacherous waters to come to Taiwan hundreds of years ago?*
>
> *Do they know?*
>
> *What was the reason that centuries ago, the forebears of the Taiwanese people suffered, worked diligently, saved every penny, and struggled the way they did?*
>
> *Do they know?*
>
> *They want to field their presidential candidate, but do they know the true aspirations of the Taiwanese people? Do they understand the suffering of the Taiwanese people? Do they know that in casting off Chinese rule, Taiwan has emerged an oceanic state?*

This was unusually heady stuff to be discussing in public, let alone on live television in front of the entire country. Although he would not win the election, Peng is credited with putting Taiwanese self-determination front and centre in the new democratic era. Another politician might have struggled to be so bold, direct and unapologetically Taiwanese on that stage, but not Peng Ming-min. He spoke not only from the heart, but from his own lived experience.

Peng was born in 1923 with the Taiwanese name Phêⁿ Bêng-bín in the central town of Taiko (present-day Dajia), under Japanese rule. He would be one of the many Taiwanese who studied in Japan during the last days of the Second World War, when American bombing raids devastated the doomed empire.

By April 1945, one of the few big cities that had been more or less spared from the bombing was Nagasaki. Peng had already abandoned his studies at Imperial Tokyo University as the city was bombarded, and so decided to join his brother in a village 30 kilometres from Nagasaki.

After several third-class train rides, he arrived in Nagasaki in the late afternoon, deciding to spend the night in the city after buying a ticket for a ferry across the bay the next day, which would get him most of the way to his brother's home. After a good night's sleep, he went to the ferry terminal and boarded his vessel for a short skip to the other side. He was filled with optimism, as the weather was beautiful and the waters calm. As he wrote in his 1972 autobiography, *A Taste of Freedom: Memoirs of a Taiwanese Independence Leader*, his tranquil feeling would prove fleeting:

> *As I went aboard and looked about for a sheltered spot, I heard overhead the peculiar whistling sound of a plane gliding down. It banked and began a steep climb with a sudden roar of motors. An instant later I was knocked unconscious to the deck by a tremendous explosion.*
>
> *When I regained my senses and opened my eyes, I was in the midst of screaming confusion and a scene of horror. I was covered with blood, and the deck was awash with it, and strewn with bodies, the shattered parts of bodies, and people writhing, moaning, and struggling to drag themselves away. I tried to rise, and found to my disbelief and horror that my left arm*

> had been torn off at the shoulder and was hanging there by a few tendons and a shred of skin. The shattered bones were exposed, and blood was pouring out. 'This is the end,' I thought, 'I am dying here, and my parents and brother don't even know I am here.'[6]

With his right hand, Peng grabbed his cold, lifeless left forearm and, as blood streamed from his arm and head, somehow managed to get himself to a clinic, where he was given a bed. Doctors were overwhelmed by the carnage and were not attending to Peng as he began to fade. Then, a miracle: one of the doctors was Taiwanese and happened to be the best friend of Peng's brother. Upon recognizing Peng, the doctor gave him life-saving drugs and secured transport to a nearby hospital, where a surgeon amputated Peng's left arm at the shoulder.

Peng made it to his brother's home, where he convalesced through the summer. On 18 June, the Allies captured Okinawa – a major blow to Japanese morale. On 7 August, Nagasaki newspapers announced: 'Yesterday Hiroshima was bombed. The Americans have used a new weapon. There was considerable damage.'

The understated announcement of the world's first atomic bombing made little impression on Peng at the time. The world's second atomic bombing would change his life for ever. As he recounted in his book:

> Three days later I was indoors, glancing through the newspaper when I heard the drone of a plane overhead. Suddenly there was a blinding light, as if a huge photo-flashbulb had been triggered in the room. This was followed instantly by a tremendous metallic clanging sound, as if the whole earth had been hit by a gigantic hammer. Our house shook violently. Something prompted me to cry out in Formosan 'What is it?' as I looked out to see an enormous black cloud over Nagasaki. Then the great white mushroom rose above it. Later there was a sudden light shower in our garden, falling out of a clear sky [. . .]

> That afternoon we heard that Nagasaki had been destroyed, zenmatsu ('obliterated'). The Americans had used their new weapon again. It was rumored that everyone in Nagasaki was dead. When my brother returned, late that night, he was in a state of shock and nausea. He could barely speak and had to struggle for words with which to tell us what he had seen. The city as we knew it was gone.[7]

Days later, Japanese subjects of Emperor Hirohito heard his voice for the first time. Over the radio, he asked them to 'bear the unbearable' and accept defeat. Peng and his brother decided to return to Taiwan, where Chiang Kai-shek's KMT army would also soon arrive.

Under the KMT, Peng witnessed the deterioration of social order, the economy and the general mood in Taiwan. While Peng was in Taipei studying for his bachelor's degree in law at National Taiwan University, the 228 uprising and massacres took place. His father was in Kaohsiung, where he had been named chairman of the local Settlement Committee. Following the arrival of KMT reinforcements in Keelung and Kaohsiung and the ensuing slaughter, Peng's father and many of his colleagues were rounded up and on the brink of being executed. It was none other than Peng Meng-chi, the reviled Butcher of Kaohsiung, who spared Peng's father just before he was executed, saying he knew him and knew that he was a good man. Peng's father returned home, unable to eat for days. Any connections the elder Peng felt to China or Chineseness were completely severed.

Peng would focus on continuing his studies, earning his master's degree in law at McGill University in Montreal in 1953, then his doctorate in law at the University of Paris. He focused on the emerging field of international air law, with his writing gaining attention both at home and abroad. His work caught the eye of Chiang Kai-shek, who named him an adviser to the ROC's UN delegation in 1960.

Peng's time in New York, far from the oppressive environment of martial law, opened up new avenues of free thought and expression. He eventually began to lead what he termed his 'double life', as he served the KMT party-state in Taiwan while also embracing his Taiwanese identity in New York. Peng would return to Taipei in 1962 to discover that he was unable to suppress his growing political consciousness. He and two of his students at National Taiwan University composed a manifesto – 'A Declaration of Taiwanese Self-Salvation' – calling for the end of plans to retake China, a new constitution and the acknowledgement that the two sides of the Taiwan Strait were separate countries. Their plans were thwarted when a printer they had contacted tipped off the police and, in 1964, they were arrested and interrogated repeatedly about who their backers were. Their interrogators insinuated that the US might be supporting their activities, underscoring KMT paranoia about losing American support.[8]

Peng and his students were convicted of sedition by a military court. One year into Peng's seven-year prison sentence, US diplomatic pressure led the Generalissimo to place his former adviser under house arrest instead. Peng would remain under house arrest until 1970 when, with Amnesty International's assistance, he escaped and fled to Sweden using a counterfeit Japanese passport, eventually making his way to the US.

There, Peng used his platform to advocate for Taiwan's democratic struggle. One of the first things he did after arriving in the US was pen an opinion piece in *The New York Times* after the transfer of the China seat in the UN from the ROC to the PRC in 1971.

The Chinese, he wrote, 'must learn to distinguish ethnic origin and culture from politics and law, and to discard their archaic obsession to claim anyone of Chinese ancestry as legally Chinese, however far removed from China.'

For the Taiwanese, the most important issue is not independence, but self-determination for its people, he wrote. The people of Taiwan, he continued, 'want to live in the most friendly association with the Chinese people, and would spare no effort to establish the closest economic, commercial and even political ties with China'.⁹

While in the US, he secured a professorship at the University of Michigan at Ann Arbor and wrote *A Taste of Freedom*. Like Su Beng's *Taiwan's 400-Year History*, Peng's book had a major impact on the global Taiwanese diaspora. Originally published seven years after George Kerr's *Formosa Betrayed*, it is also one of the first English-language books to introduce a global audience to Taiwan's struggle for democracy.

Peng would go on to co-found the Formosan Association for Public Affairs (FAPA) in 1981. This Washington-based organization became a key player in lobbying for Taiwanese interests in the US. FAPA is still very active in advocating for Taiwan today, and has one of the largest bases of bipartisan support in Congress. Some Taiwan-related bills that have been signed into law were originally drafted by FAPA. One example of this is the Taiwan Travel Act, which President Donald Trump signed into law in 2019 after the Senate passed it on 28 February ('228') 2018. The act encourages visits by American officials at all levels to Taiwan while pushing for dignified treatment for Taiwanese officials in the US. The act was originally proposed by FAPA in 2005, with the organization lobbying for it for fourteen years, eventually achieving success.[10]

After advocating for Taiwan in the US for more than two decades and with democracy taking root in Taiwan, Peng finally flew back to his homeland in 1992. He was greeted by a crowd of around a thousand supporters. Two years later, he joined the DPP, which chose him as its first legally recognized presidential candidate. His running mate was Frank Hsieh, the DPP co-founder who had worked on the Henry Liu case twelve years earlier.

Peng and Hsieh would finish second to Lee Teng-hui, with less than half of the votes of the KMT incumbent. It would be a solid first showing for the DPP and help pave the way for Chen Shui-bian's victory in 2000. In recognition of Peng's contributions to Taiwanese self-determination, Chen would name him a presidential adviser.

For many politically minded Taiwanese today, Peng's vision of a democratic and sovereign Taiwan is still an inspiration. Kolas Yotaka, the former presidential spokeswoman and legislator, told me that she was deeply impacted by Peng's 1995 speeches. She said she hears echoes of Peng's words in speeches made by modern politicians who respect the sacrifices he made: 'Peng showed Taiwanese by example that, even though we had been under dictatorship for half a century, democracy was still within our reach.'

CHAPTER 8

The Democracy Experiment

Bright, winsome and well-heeled, Tainan native Wu Shu-chen was the daughter of a paediatrician father and a mother from a wealthy family. In her twenties, Wu was pressed by her parents to marry a doctor who could carry on the family practice. While studying law in Taipei in the 1970s, however, Wu fell in love with a former middle school classmate who was also training to become a lawyer. The son of a poor farmer, he had grown up in a rural part of Tainan, in a home that was made of mud bricks. His name was Chen Shui-bian.

Young and idealistic about Taiwan's future, Wu and Chen married in 1975, to her parents' dismay. Photos from their early years together show the couple smiling, aglow with young love. Chen donned thick black-framed glasses and Wu had her long black hair pulled back. They would start their lives together at a time of momentous changes in Taiwan – changes that would eventually engulf their lives.

Following the Kaohsiung Incident of December 1979, Wu convinced her husband to join the defence team of the Kaohsiung Eight. At only thirty years old, he was the youngest member of the team. During the trial, Chen argued passionately against the use of coerced confessions as evidence. While operating under the authoritarian rule of the KMT party-state, such an argument was not only bold but potentially dangerous.

'Your honour, please investigate the legality of the confessions,' Chen said to the presiding judge. 'If they were extracted by torture, not given of their own free will, the charges brought against the accused are clearly dubious and we can stop ourselves from wasting any more time debating the facts of this case.'[1]

The six men and two women of the Kaohsiung Eight were on trial for sedition against the ROC – a charge which could lead to the death penalty. The defence team's primary and rather ambitious goal was to prevent any of their clients from being executed. When the verdict was returned only one defendant, the defiant Shih Ming-teh, was sentenced to life in prison. The other seven received prison terms ranging from ten to fourteen years. Considering the potential for more severe punishments, the defence team viewed the avoidance of capital punishment for all defendants as a significant victory.

'At that time, anyone tried by a military court was supposed to get the death sentence, unless granted clemency by the president,' said Chen Chi-sen, who led the defence team of fifteen lawyers. 'We couldn't let them die – it was our only goal to keep these people alive. For that reason, we felt we had won the case.

'We had to overturn their confessions, which were seen as very powerful evidence,' Chen Chi-sen continued. 'But we also had to restrain ourselves from going too far. Otherwise, our efforts would have been in vain.'[2]

The trial captivated the nation. The two major dailies of the time, the *China Times* and *United Daily News*, flew off the shelves during the nine-day trial. The bravery of the Kaohsiung Eight and the competence of their defence team inspired millions of Taiwanese to contemplate what a post-martial law Taiwan might look like – and who might be fit to lead it. Between the eight defendants and their fifteen lawyers were multiple future DPP mayors, legislators and ministers, plus one premier, one vice president, one president and multiple DPP chairs.

Following his notable defence in the trial, Chen Shui-bian, affectionately known as A-bian by his supporters, became more active in the Tangwai movement. He also built a successful legal practice focusing on maritime and commercial law. In 1981, he was elected a Taipei city councillor and in November 1985 he ran for magistrate of his home county of Tainan. Despite narrowly losing to his KMT opponent, Chen, alongside his wife Wu Shu-chen, travelled across Tainan to thank voters for their support.

The couple made a stop in Tainan's Guanmiao District, known for growing the country's most succulent pineapples. Outside a large local temple, a small farm truck struck Wu, running over her. The vehicle then backed over her, and ran over her again. Wu miraculously survived but was left paralysed from the waist down. The driver of the truck was detained and quickly released without charges – the authorities deemed it an accident. Chen and many of his supporters believe that the KMT had orchestrated Wu's attempted murder as a form of political reprisal.

Only months later, the couple would face further adversity. In 1986, Chen was imprisoned for libel after writing an article criticizing Elmer Fung, secretary to Chiang Ching-kuo. With her husband in prison, rather than be cowed, the wheelchair-bound Wu took the fight to the KMT by running for and winning a legislative seat representing Taipei for the DPP. Chen would be released in early 1987, after which he served as Wu's chief counsel.

Wu would only serve one term as a legislator, shifting the couple's focus entirely to Chen's political career. Chen won terms in the Legislative Yuan in 1989 and again in 1992, where he further consolidated his reputation as a formidable opponent to the KMT. In one famous moment in 1993, comments made by Chen so upset the KMT legislator Han Kuo-yu that he flipped a table over on Chen in the legislature. Later the same day, he sucker-punched a seated Chen in the left ear. Chen was hospitalized for three days, garnering sympathy for once again having been victimized by the KMT.

Chen kept moving up the political ladder, narrowly winning a tight three-way race for Taipei mayor in 1994. The DPP had only existed for eight years when Chen took office, so the party's pool of members with experience in governance was quite shallow. As a result, many of Mayor Chen's staff outside his trusted inner circle of DPP comrades were KMT members. Despite this, Chen's tenure as mayor was marked by significant achievements in reducing crime, improving public infrastructure and elevating Taiwan's Indigenous history. One prominent example was the renaming of Chieh-shou Boulevard ('Long Live Chiang Kai-shek Boulevard') to Ketagalan Boulevard, after one of the several Indigenous tribes native to northern Taiwan. Additionally, he designated the former radio station where the call for popular revolt that led to the 228 uprising took place as a site of historical significance. His administration also chose to make the building the home of the Taipei 228 Memorial Museum.

In a 1997 interview with *The Christian Science Monitor*, Chen explained the significance of addressing 228 and removing the Chiangs' names and images from the public sphere in democratic Taiwan.

'There were so many taboos here under martial law that even after it formally ended, many people were still trapped by invisible barriers on what they could think and do,' he said.[3]

Chen's approach to governance was unlike anything Taiwan had seen up to that point. While his KMT counterparts were generally viewed as aloof and out-of-touch, Chen would answer audience questions on live radio and television programmes. He set up town hall-style events where he would meet with Taipei residents, addressing their concerns and fostering a more inclusive political discourse. As mayor of the capital, Chen did not shy away from addressing the complex and contentious issue of relations with

China. He boldly advocated for the cessation of the ban on direct trade and transportation links with China, while also calling for resumption of cross-strait talks that had been frozen since 1995.

Despite the popularity of his social and economic policies, and his mayorship in general, in 1998 he lost his re-election bid to Ma Ying-jeou of the KMT. This defeat would not be the end of Chen's political journey, however. Eyeing the KMT's declining popularity amid corruption scandals and internal divisions, Chen announced his candidacy for the 2000 presidential election.

The political environment during the 2000 election was ripe for change. The KMT was bogged down by its association with *heijin*, or 'black gold' – the intertwining of organized crime and money politics in Taiwan. Chen's campaign was further buoyed by the split in the blue, pro-KMT camp between Lien Chan and James Soong. Chen and his running mate Annette Lu squeezed through that split. History had been made. Chen Shui-bian became the first president in the ROC's fifty-five years in Taiwan to not be a member of the KMT.

Beyond the significance of the DPP winning the presidency in Taiwan's second-ever election, Chen's victory also represented an important generational shift. Lee Teng-hui's generation, which had grown up as Japanese colonial subjects, was passing the baton to Taiwanese who had grown up as Chinese colonial subjects.

In the lead-up to Chen taking office in early 2000, the KMT legislator Su Chi made a significant announcement. He claimed that during informal talks in 1992 between the ROC and PRC, an understanding that would become known as the 1992 Consensus had been reached. The consensus, according to Su, was an agreement that Taiwan and China are part of the same country, allowing for each party's own interpretation of what that meant.

However, the announcement of this consensus was not made

until eight years later, raising doubts among many Taiwanese about its authenticity. These suspicions were rekindled in 2006, when Su Chi admitted to fabricating the 1992 Consensus as an attempt to stabilize cross-strait ties before Chen took office.[4] This admission was made one day after former president Lee Teng-hui dismissed the consensus as a fabrication, accusing Su of being a 'little monkey boy trying to rewrite history'.

Despite Su's admission, Xi Jinping's China insists that the 1992 Consensus is real, maintaining that Taiwan belongs to the PRC. The meetings said to have led to the consensus were conducted between two unofficial government organizations: the Straits Exchange Foundation (SEF) representing Taiwan, and the Association for Relations Across the Taiwan Strait (ARATS) from China. However, in October 2024, the SEF Secretary-General Luo Wen-jia told Taiwanese media that no record of the 1992 Consensus could be found in the SEF's archives or its correspondence with ARATS in 1992.[5] Regardless, Beijing still uses this fabrication, originally intended to constrain Chen's presidency, to portray the DPP as 'separatists'.

Upon taking office, Chen Shui-bian once again faced the challenge of not having enough experienced members from his party to fill his administration. Once more, he was forced to appoint KMT loyalists to key posts. This included General Tang Fei, the incumbent defence minister, as his first premier, a role responsible for managing day-to-day government operations, allowing the president to focus on defence and foreign relations. Chen's inclusion of a general in his cabinet may have also lessened the likelihood of a military coup against him.

Chen began his presidency with high popularity, but his administration encountered difficulties in passing legislation due to the KMT's parliamentary majority. Such an impasse was unprecedented, given that the KMT had always previously controlled both the executive and legislative branches. Moreover, the Taiwanese stock market suffered greatly, losing half its value in one year. Tensions escalated between Chen and the KMT-controlled Legislative Yuan over an unfinished nuclear power plant. The DPP had opposed nuclear energy since its establishment, whereas the KMT viewed it as essential for maintaining Taiwan's economic growth. Less than five months into office, General Tang Fei resigned as premier over the nuclear plant issue.

Chen did make significant efforts to navigate the complex political terrain with the KMT, while also enhancing Taiwan's international standing and fostering cross-strait relations with China. Chen's administration also undertook strategic initiatives to bolster Taiwan's global profile, especially within the US through visits to key cities – Los Angeles, Houston and New York – where he engaged with local officials and members of Congress, and received honours such as the key to the city of Houston. One of the most significant items of legislation that Chen was able to get through the KMT-controlled parliament was a referendum law with a clause that permitted votes on sovereignty issues and 'issues of national security concern' in the case of an imminent external threat. This opened the possibility of a vote on establishing a Taiwanese state in the face of a looming Chinese attack.

At the same time, the Chen administration made overtures towards China, facilitating the establishment of the 'mini three links'. These links permitted postal, transport and commercial connections, but were initially limited to residents of the Taipei-controlled islands of Kinmen and Matsu just off the Chinese coast, or for *Taishang* – Taiwanese businesspeople with enterprises in China. This initiative marked a cautious yet significant step towards easing travel and commercial restrictions between the two sides of the Taiwan Strait.

Yet as these conciliatory moves towards China were made, the events of 11 September 2001 in the US also brought the US and China closer than ever, in turn leaving Taiwan more vulnerable. The

Bush administration perceived China, under the leadership of Hu Jintao, as an ally in the US's so-called War on Terror, and believing the Chinese could use their leverage to curb North Korea's nuclear ambitions. Instead, Hu used the War on Terror as an excuse to intensify repression of Muslim Uyghurs, while doing nothing to further Bush's aim of ending Pyongyang's nuclear programme.

Taiwan's early democratization coincided with the astonishing growth of the US–China economic relationship. Between 1989 and 2003, US exports to China quadrupled from $5.8 billion to $26.1 billion. Chinese exports to the US also shot up twelvefold, despite starting from twice as large a base. China's trade surplus with the US spiked twentyfold, hitting $119.5 billion. An emboldened China had strengthened its position globally, a fact that formed part of the backdrop to the 2004 Taiwanese election campaign.

The Elections of 2004 and 2008

In late November 2003, with the next Taiwanese election on the horizon, Chen Shui-bian announced at a campaign rally that China had 496 ballistic missiles aimed at Taiwan. He proposed holding a 'defensive referendum' alongside the presidential election in March 2004, igniting a chain of diplomatic and political manoeuvres.[6]

China, viewing the proposed referendum as a step towards Taiwanese independence, issued vague threats of a strong reaction should the referendum take place. With Beijing's increasing political and economic leverage over Washington, the situation put strain on US–China relations. The Bush administration, prioritizing its strategic interests, was clearly displeased with Chen's referendum proposal.

The diplomatic tensions came to a head less than two weeks after Chen's call for the referendum when Chinese head of government Wen Jiabao met with Bush at the White House. During this meeting, Wen hoped the American president might say exactly what Beijing wanted to hear. Bush delivered.

'We oppose any unilateral decision by either China or Taiwan to change the status quo,' Bush told reporters, 'and the comments and actions made by the leader of Taiwan indicate that he may be willing to make decisions unilaterally, to change the status quo, which we oppose.'[7]

Of China, Bush said, 'We are working together in the War on Terror,' calling the countries 'partners in diplomacy' and 'full members of a world trading system'.

Bush's remarks, made while standing alongside an obviously satisfied Wen, sidestepped directly acknowledging Chen as Taiwan's president. This was not helpful to Chen's re-election hopes. As Chen and Annette Lu faced declining popularity in the final months of their campaign, the KMT's Lien Chan and his running mate James Soong of the People First Party had a serious chance of defeating the DPP incumbents.

In January 2004, Chen spoke to Taiwanese voters in a televised address, in which he announced the wording of the two questions that would be on the ballot of the defence referendum. The first question was:

> The people of Taiwan demand that the Taiwan Strait issue be resolved through peaceful means. Should the Communist Party of China refuse to withdraw the missiles it has targeted at Taiwan and to openly renounce the use of force against us, would you agree that the Government should acquire more advanced anti-missile weapons to strengthen Taiwan's self-defence capabilities?

And the second question:

> Would you agree that our Government should engage in negotiation with the Communist Party of China on the establishment of a 'peace and

stability' framework for cross-strait interactions in order to build consensus and for the welfare of the peoples on both sides?

During his televised speech, Chen reaffirmed the pledge he made at his May 2000 inauguration. He vowed not to declare independence, change the nation's official name, alter the ROC constitution in a manner that formalized Taiwan and China as separate entities, or initiate a referendum on independence or unification. There was one caveat for all these pledges, however: China must not intend to use military force against Taiwan.

On a cool afternoon in 2004, Taiwan's entire west coast became a powerful protest against the Chinese missile threat. On 28 February, approximately two million Taiwanese – one-tenth of the population at the time – joined hands to form a 300-mile human chain snaking from Keelung to Kaohsiung. Facing the strait – and China – participants young and old chanted 'Cherish peace! Oppose missiles!' At the symbolic time of 2.28 p.m., everyone pivoted eastwards, turning their backs on China.

Chen and former president Lee Teng-hui, who organized the demonstration, joined hands in Miaoli County, home to large Hakka and Indigenous populations. Chen described the sight as a 'democratic Great Wall'. It was a clear statement of Taiwan's desire for peace and opposition to military intimidation. 'This is an historic attempt by the people to show the world their love for Taiwan and the power that people have to write history with their own hands,' he said at the rally.[8]

Lee modelled the demonstration on the Baltic Way, a 1989 peaceful protest by the Baltic states of Lithuania, Latvia and Estonia that would lead all three states to declare independence from the Soviet Union two years later. Today, the Baltic states are democratic and free, as well as members of NATO and the EU.

The eighty-one-year-old Lee declared the massive human chain to be 'a victory for the people of Taiwan' and 'the people's affirmation of Taiwan's national identity and a rejection of China's missile threats'.

An ebullient Chen said the rally was just the beginning. 'We must continue our effort and cast referendum ballots on election day to maintain Taiwan's security,' he said.

The rally and the impending referendum were met with resistance from the KMT and smaller blue parties such as the People First Party and New Party. They argued against participation and urged their supporters to boycott the referendum on election day. This strategy, combined with a narrow lead for the Lien–Soong ticket in the polls, set the stage for an intense electoral showdown.

The day before polls opened, Chen Shui-bian and Vice President Annette Lu were wrapping up their campaigning. At a crowded rally in downtown Tainan, they were standing in the back of a slow-moving red jeep, from which they waved and gave thumbs-up signs to supporters. The jeep was escorted by nine police on motorcycles as the crowd cheered and fireworks exploded. Amid the popping firecrackers, someone fired two bullets at the candidates. One lodged itself in Chen's stomach, the other hit Lu's right knee. Both were rushed to hospital and discharged the same day. All political rallies scheduled for that night across Taiwan were cancelled.[9]

The shooting had a profound impact on the electorate, influencing public sentiment and arguably swaying voter sympathy towards Chen and Lu. They would eke out a re-election victory the following day. Despite the more than 80 per cent voter turnout, the margin was razor thin, even compared to Chen's previous victories, although he gained a significant number of votes from his 2000 victory. Chen and Lu won 50.11 per cent of the vote, with Lien and Soong garnering 49.89 per cent. With nearly sixteen million casting their vote, the election was decided by a margin of less than 30,000 ballots.

Following the election, the KMT and its allies raised doubts about the legitimacy of the shooting, suggesting that they had been staged to garner sympathy votes – a claim echoed by Chinese state media. However, investigations later ruled that the shootings had been real. The shootings aside, Chen's referenda were impacted by the blue boycott, failing to pass due to not meeting the required 50 per cent voter-turnout threshold. Nonetheless, there was overwhelming support for both propositions among those who did vote, with more than 91 per cent in favour.

While Chen may have been victorious, his second four-year term as president would not be easy. It is best remembered in Taiwan for the allegations of corruption against him and his wife, Wu Shu-chen, which cast long shadows over his tenure. It also set the stage for Ma Ying-jeou's 2008 election as president.

From the outset of Chen's presidency in 2000, he faced relentless opposition from a KMT that still controlled the Legislative Yuan. The KMT embarked on a determined effort to remove him from office, including passing legislation making it easier to impeach the president, and KMT members attempting unsuccessfully to recall him. It wasn't until 2006 that the KMT found a political figurehead to rally behind who could seriously challenge Chen. Ironically, that leader was DPP doyen Shih Ming-teh, one of the Kaohsiung Eight, whom Chen had defended in 1980.

Shih was born in Kaohsiung in January 1941, towards the end of Japan's half-century of colonization of Taiwan. The 1945 bombing of his hometown by the US was one of his earliest memories. When he was just a year and a half older in 1947, the violence of 228 would be seared into his mind when he witnessed KMT soldiers gunning down university students outside Kaohsiung's rail station.[10]

Shih signed up for military school in 1959, telling his mother he did so to learn how to lead an insurrection against the KMT. Tensions were high due to Communist attempts to take the numerous islands just off the Chinese coast that were still under KMT control. He graduated in 1961 and was sent to one of those island groups – Quemoy (today Kinmen), just 5 kilometres off China's Fujian coast.

Having witnessed the end of the Second World War and the beginning of martial law in Taiwan as a boy, twenty-year-old Shih found himself an artillery officer serving in the Chinese Civil War. Two years later, in 1962, he was arrested by military police for involvement in a secret alliance of independence-minded youth. More than thirty others were also arrested, mostly military academy and university students. When Shih was interrogated, military police beat his mouth, shattering most of his teeth. In 1964, twenty-three-year-old Shih was sentenced to life in prison.

Following Chiang Kai-shek's death in 1975, Chiang Ching-kuo demonstrated leniency to the thousands of political prisoners in Taiwan. Shih's life sentence was commuted to fifteen years and he was freed in July 1977. A free man once more, the determined Shih jumped into the Tangwai movement, despite enduring torture in prison and the constant threat of death from the KMT. Knowing that he was risking death, Shih didn't use telephones out of concern that they were tapped. That meant spending much of his time relaying messages and meeting fellow Tangwai members by foot.

'I could see that he was working like a man on fire to challenge the authoritarian rule,' said Linda Gail Arrigo, an American scholar and democracy activist, who married Shih in 1978. 'He expected to die in prison – by execution.'[11]

Following the clashes between police and protestors in the Tangwai-led demonstration for human rights in Kaohsiung in December 1979, Shih went on the run, eluding arrest for a month before finally being captured. His arrest photo shows his jaw wrapped in bandages – the result of a botched attempt to receive plastic surgery from a dentist. Undeterred, Shih used hunger strikes to protest within the prison walls. He was force-fed through a nasogastric tube more than 3,000 times in the 1980s.

His determination began to attract global attention, at a time when the KMT was under growing pressure to end martial law. In 1984, the Polish democracy activist Lech Wałęsa nominated Shih for the Nobel Peace Prize. A former political prisoner himself, Wałęsa had won the prize the year before.

Following the end of martial law in 1987, both Chiang Ching-kuo and, after Chiang's death, President Lee Teng-hui offered Shih an early release. He rejected both, choosing to stay in prison until he was fully exonerated. That moment came in 1990, when the verdicts in the Kaohsiung Eight trial were overturned. After a decade in prison, Shih Ming-teh emerged into a radically different Taiwan. He was, of course, a hero. The Tangwai movement of which he had been the spiritual leader was now the DPP — the first legal opposition party under any of Taiwan's colonial governments.

Shih Ming-teh would gradually morph from a leading figure within the DPP to the head of a largely blue movement against President Chen Shui-bian. Starting with his key victories in Taiwan's nascent stages of direct democracy — first as a legislator for Tainan in 1992, then rising to the position of DPP chair — Shih played a pivotal role in shaping the party's early trajectory. But after Lee Teng-hui of the KMT's heavy defeat of the DPP candidate Peng Min-min in the 1996 presidential election, the DPP underwent a period of introspection and recalibration, leading to Shih's eventual estrangement.

After the 1996 election, Shih Ming-teh resigned as party chair. At a DPP leadership meeting in early 2000, the lawyers would wrest the reins of the party from the activists, moderate its stance towards independence and emphasize strengthening Taiwan's economic ties with China, a strategy instrumental in securing Chen Shui-bian's presidential victory in 2000. Despite Chen's offer of a senior position within his administration, Shih chose to instead distance himself from the political limelight.

The watershed moment of Shih's political career came in 2006, when he orchestrated mass protests against President Chen, accusing him and his administration of corruption. Chen's son-in-law was under investigation for insider trading. His wife, Wu Shu-chen, was accused of accepting large amounts of gift certificates from a department store chain that was lobbying the government for a change of ownership. Investigators were also examining whether Chen and Wu had properly accounted for a presidential fund reserved for diplomacy. Shih dubbed his movement 'Million Voices Against Corruption, Chen Must Go', but it would informally be known as the redshirt protests, due to the seas of red-clad supporters. Notably, the overwhelming majority of the protestors were supporters of the KMT. For many members of the DPP, this was a bitter betrayal by Shih, tarnishing everything he had done before for the Tangwai movement.

Chen refused to step down and, despite being immune to questioning while president, spoke with investigators. The secretary-general of the presidential office, Mark Chen, noted that the same KMT that imprisoned Shih was now the main force rallying behind him against a president it had been trying to depose for years. 'I hope he doesn't forget those are the guys who tried to kill him,' Mark Chen said, 'and now they support him like he is a god.'[12]

Shih's campaign put an indelible stain on Chen's presidency, one which the KMT – no stranger to corruption itself – still uses to attack the DPP today. The failed movement would also lead to the end of Shih's political relevance and his banishment from the green camp.

It would be even more ruinous for Chen. After stepping down in 2008, he was immediately placed under investigation and was not allowed to leave the country. His wife Wu Shu-chen had been indicted in 2006 – now without presidential immunity and the KMT back in charge, it would be Chen's turn, and he would be shown no mercy. In 2009, Chen was found guilty of multiple financial crimes. He was sentenced to life in prison, and thrown into a small cell with one cellmate. In 2010, the ROC High Court reduced his sentence to twenty years. His mental and physical health deteriorated, and in 2012 he had a stroke, after which he was diagnosed with sleep apnoea and severe depression. In 2013, Chen attempted suicide.

He was discharged on medical parole in 2015 but remains a second-class citizen with limited rights. Most notably, he is forbidden from engaging in any public political speech. Additionally, his medical parole doesn't count towards his twenty years, so around fourteen years of his revised sentence remains unserved.

It's difficult not to find Chen's punishment excessive. Especially given that his conviction for embezzlement of state funds was overturned by Taipei District Court in 2010, leading to his reduced sentence. The other charges include taking bribes and attempting to hide funds abroad, and while it was found that former First Lady Wu received large sums of money from people and organizations with interests before the state, no direct case of quid pro quo was proven.[13]

Taiwan-based legal consultant and political commentator Michael Fahey was a journalist at Formosa TV covering the red-shirt protests in 2006. While there were certainly questions about whether Chen Shui-bian had violated some laws, there were also questions about the fairness of the proceedings that took his freedom, he said.

'I think it's probably safe to say that most people in Taiwan think he was corrupt, but there have always been questions about whether the supposed bribes were actually campaign contributions, and whether politics prevailed over due process in the state's prosecutions,' Fahey told me. 'It's very complicated, and so are the cases and legal doctrines that were used.' When I pushed Fahey on whether or not he thought Chen was guilty, he shrugged.

'In short, I don't know. I don't know anyone who I think really does. Shih Ming-teh used his enormous moral authority to persuade the public that Chen was corrupt. I firmly believe he did that because he thought he deserved to be president and his jealousy, resentment or craving for attention was manipulated.'

With the 2008 election of the KMT's Ma Ying-jeou over the DPP's Frank Hsieh, the KMT was back in charge of the executive and legislative branches and the DPP was in shambles. Looking at the mess left by the unravelling of the Chen presidency, the mostly male party leadership chose Tsai Ing-wen as party chair. At fifty-one, she was the youngest chair in the party's short history, as well as the first woman to hold the position. As DPP members lamented that the party wouldn't return to power for another two decades, Tsai commenced her rebuilding work.

Meanwhile, Ma Ying-jeou, given a mandate to lead by disaffected voters, got to work on bringing China and Taiwan closer. He had campaigned on a platform of letting Taiwanese decide their own path forward. But his style once in charge might be best summarized as 'Make Taiwan China Again'. During his eight years as president, he would bring Taiwan closer than ever to the PRC, while rejecting Taiwanese sovereignty through his public endorsement of the 1992 Consensus. As China continued to build up its military capabilities with an eye on eventually taking Taiwan, Ma reduced the mandatory conscription time for young men from one year to four months, and moved to an all-volunteer force, weakening Taiwan's military preparedness while neglecting to respond to the massive build-up across the strait. Instead, Ma prioritized the flow of Chinese money and people into Taiwan.

In 2005, largely as a result of Chen's re-election, the KMT and the CCP had normalized party-to-party relations. Many Taiwanese saw ties with their booming neighbour primarily through an economic lens and Beijing's military impotence in the face of the US Seventh Fleet in 1996 was still fresh in many minds. During the reforms of the Chen years, Taiwanese civil society – the non-partisan inheritors of the Tangwai legacy – became less engaged in politics and more focused on their livelihoods. The Ma era, however, would bring them back to the streets.

Wild Strawberries and Sunflowers: Lin Fei-fan

A few weeks before Taiwan's 2024 presidential and parliamentary election, I sat down with Lin Fei-fan at his office at the west edge of Taipei's Zhongzheng District, where much of the city's old Japanese-era administrative buildings are located.

Lin, then thirty-five, was born months after the end of martial law. He is a native of Tainan, where he studied political science at National Chung Kung University. I had originally heard of Lin as one of the leaders of the Sunflower Movement in 2014. Like many politicians in Taiwan before him, he has since moved from street protests and occupations to party politics.

When I first met Lin in 2018, he was back in Taipei after obtaining a master's degree in comparative politics at the London School of Economics. At that time, he was hosting a pro-democracy activist group with members from Taiwan, Hong Kong, Thailand, Cambodia, Vietnam, Japan and elsewhere around the region. Taiwan was a rare bright spot for human rights in Asia and the following year would become Asia's first country to legalize same-sex marriage. The pan-Asian pro-democracy group had decided to meet in Taiwan because it was the only country in East or Southeast Asia where all members were certain they would be allowed to enter. The Hong Kong activist Joshua Wong, now a political prisoner, had already been barred from entering Thailand in late 2016.

The erosion of freedoms in Hong Kong would lead to Taiwan becoming Asia's bastion of free expression just three decades after the end of martial law. One member of the group from Hong Kong, Johnson Yeung, told me after the group's meeting in Taipei: 'I felt really glad that Taiwan could still provide this kind of safe space for activists from different parts of Asia.'

By late 2023, Lin was the DPP's deputy secretary-general, a rather senior position for someone in their thirties. Months later, in 2024, the incoming president, Lai Ching-te, would appoint him as deputy secretary-general of Taiwan's National Security Council.

Wearing round wireframe glasses and an untucked blue flannel shirt, Lin was, as in my previous meetings with him, a smart and serious person possessing a relaxed, friendly demeanour. As an elementary school student, Lin would go to DPP rallies with his father, an ardent supporter of the party. As a fifth-grader attending rallies in the run-up to Chen Shui-bian's history-making victory in 2000, Lin said that he 'was starting to have a sense of what politics was'. A major theme of the 2000 election was Taiwanese versus Chinese identity, with Lin telling me that despite being told in the classroom that both he and Taiwan were inherently Chinese, 'at that time, I got the sense that I was Taiwanese'.

Eight years later, a twenty-year-old Lin found himself at his first demonstration.

'When the KMT came back in 2008 you could really feel the conflict between the two identities and how you understand the future of Taiwan,' Lin Fei-fan recalled. 'Should we move into China's orbit, or should we deepen our cooperation with other nations?'

From the outset, the new president Ma Ying-jeou made clear that he was intent on forging closer ties with China. Chen Shui-bian's treatment was viewed by many in the DPP and others in the green camp as a chilling message that Ma would return to the old days of KMT authoritarian rule. Other warning signs would soon follow. Half a year into his first term, in November 2008, Ma hosted Chen Yunlin, the chairman of the Association for Relations Across the Taiwan Strait, the unofficial PRC organ for informal interactions with Taiwan. Chen was the highest-ranking

Communist official to visit Taiwan since the founding of the PRC in 1949. The extensive police presence and the pre-emptive removal of protestors during Chen Yunlin's visit alarmed many in Taiwan, especially youth and student activists. After eight years of increased democracy and freedom to express Taiwanese identity during Chen Shui-bian's presidency, most of Taiwanese civil society had either gone into government or had decreased their activity – but this visit rekindled the spirit of activism. It catalysed a wave of organization and protests, with students coordinating across various digital platforms such as Yahoo! Live, MSN Messenger and the old-school Taiwanese chat board PTT.

Lin's first foray into public activism would be a humble one. When he arrived at the front gate of his university, he said there were no more than twenty people there to protest. Along with Taipei, Tainan was one of six cities where protests took place, also including the urban centres of Hsinchu, Taichung, Chiayi and Kaohsiung. The protestors would dub themselves the Wild Strawberry Movement: the 'wild' a nod to the 1990 Wild Lily student protests and the 'strawberry' a reference to the pejorative nickname that older Taiwanese used to describe youth – the strawberry generation – because of their perceived fragility. In typical Taiwanese fashion, an anonymous internet user uploaded a song a few days into the protests, praising the movement as one of 'wild strawberries'. The song went viral, giving the Wild Strawberry Movement its name in society at large.

The mood grew heavier just a few days into the protests when, in Liberty Square, below the Chiang Kai-shek Memorial, a seventy-nine-year-old named Liu Po-yen self-immolated.[14] Liu was born in Taichung during the Japanese era and would later be part of the first wave of Taiwanese to join the KMT in 1950. He would leave the party in 2000 during the post martial-law resurgence of Taiwanese identity under the Chen administration. Before setting off northwards to Taipei during his final hours in 2008, Liu told his neighbours that he wanted to protect the Wild Strawberries. After his death, he told them, his spirit would stay in the square to protect the students.

The new chair of the DPP, Tsai Ing-wen, oversaw the organization of Liu's funeral. Among others, the funeral was attended by forty student demonstrators who declared him a martyr for the Taiwanese people. The event was one of the countless reminders of the intergenerational nature of the Taiwanese struggle for self-determination.

The students sought an apology from Ma, the resignation or firing of the directors of the national police and national security bureau, and a review of the law governing public assemblies. They got none of these, but the movement did energize the grassroots for the next big protests, which would take place four years later.

In 2012, Ma defeated Tsai Ing-wen in his re-election campaign, largely boosted by voter sentiment that closer relations with China were good for the economy. It was Tsai's first attempt at the presidency, but not her last. It wouldn't be long before activists, scholars and students – including Lin, then in Taipei working on a master's in political science at National Taiwan University – took to the streets once more. This time, it was to protest Ma's lack of regulation of the media. His administration allowed Taiwanese businessmen with extensive investments and interests in China to buy up local media and shift their editorial slant from pro-ROC to pro-PRC, and even pro-unification. In what became known as the Anti-Media Monopoly protests of 2012 and 2013, Lin found himself as the voice – and face – of a movement for the first time.

'It was a gradual process. I didn't make a decision to be a leader,' he told me. 'There's always been someone who will be put on the front line to handle the microphone, and I happened to be someone that people knew could do that.'

The Anti-Media Monopoly Movement ultimately failed in its attempt to force the government to tighten regulation on

China-connected money in Taiwan media, but it did create connections between activists that would prove crucial to perhaps the most significant protest in Taiwan's short history as a democracy – the Sunflower Movement.

Lin's experience protesting and organizing in 2012 and 2013 would prove valuable in the spring of 2014. Ma and the KMT were attempting to fast-track a trade-in-services agreement with China without adequate public scrutiny or legislative review.

Under President Ma's leadership, there was a visible shift towards fostering closer economic and political ties with China, underscored by Ma's endorsement of the 1992 Consensus. In layman's terms, this was Ma's acceptance of Taiwan as part of an entity known as China and his official rejection of Taiwanese self-determination. This stance sparked controversy in Taiwan, particularly as Ma had previously campaigned on the notion that the Taiwanese must decide Taiwan's future. By the summer of 2013, it appeared that Taiwan's future was increasingly being decided by the KMT and the CCP.

On 21 June 2013, unofficial representatives of the ROC and PRC governments signed the Cross-Strait Service Trade Agreement (CSSTA) in Shanghai. This agreement would open Taiwan's service sector – which accounted for 60 per cent of GDP – to Chinese investment. The industries affected would include financial services, telecoms, health care, hospitality and publishing. The deal alarmed many Taiwanese, who were gravely concerned about the broader implications for their economic independence and sovereignty. In the eyes of many students, the Ma administration, and more broadly the KMT, sought to sell Taiwan to China. The leader of the KMT delegation, Lin Join-sane, said after signing that the pact would 'result in a win-win situation for both sides', likening it to the 2003 Closer Economic Partnership Agreement between China and Hong Kong that accelerated Hong Kong's growing economic reliance on China.[15]

Lin Fei-fan and a cohort of activists, scholars and students – including Huang Kuo-chang and Chen Wei-ting, who had previously rallied against media monopolization – found themselves drawn once again into a struggle to preserve Taiwan's identity and self-determination. They planned to occupy Jingfu Gate, the only extant east gate of Taipei's Qing-era city wall, on 10 October 2013 – the national day of the ROC. The group's plans were apparently leaked, Lin told me. An increased police presence at the old gate led to their protest failing before it could even start.

Undeterred, they would try again in 2014. On 17 March, Lin, Huang, Chen and others including the future DPP legislator Lai Pin-yu, met one last time to plan the second attempt to occupy Jingfu Gate. After the meeting, Lin went down south to Tainan to recruit more students to join the protest. Meanwhile, Huang and others were walking home and noticed that the Legislative Yuan, Taiwan's legislature, had very few police guarding it.

When Lin returned to Taipei the following afternoon, the group notified him that the protest target had been changed from Jingfu Gate to the Legislative Yuan, and different members were assigned to different entrances around the building. Around 9 p.m. that day, Lin, his cohort and hundreds of other protestors charged the Legislative Yuan and successfully occupied it. Their actions gained widespread support from across Taiwanese society and an initially lukewarm DPP under Tsai began to champion the movement. In what became known as the Sunflower Movement, the protestors occupied the legislature for twenty-three days, successfully derailing the trade deal with China.

Days after storming the legislature, protestors attempted to occupy the Executive Yuan, which was met with a violent response by police and over a hundred arrests. After that, hundreds of thousands of Taiwanese rallied outside the presidential office building in solidarity with the protestors. On 6 April, the KMT legislative speaker, Wang Jin-pyng, part of the party's

Taiwan self-determination faction and a nemesis of Ma Ying-jeou, announced that the bill would not be put to a vote until public accountability mechanisms were in place. The protestors had won this battle. They agreed to disperse.[16] A few years later, the centenarian Taiwanese revolutionary Su Beng told me point blank: 'The Sunflower students saved Taiwan.'

Even though they were heroes to many, Lin and others had been arrested and were tried for their role in the occupation in 2017, three years after they had achieved their goal.

'I thought it was quite exciting,' Lin told me with a smile. 'I wasn't really afraid of going to jail at that time. I was devoted and not scared of any consequences. I was ready to take responsibility for my actions.'

Lin and twenty-one other Sunflower participants were all acquitted.[17] It marked a new phase of the judiciary, Lin said, as it was the first case in which the ROC judiciary supported peaceful disobedience. It was a moment in which it became clear that Taiwan's courts, once in lockstep with a Chinese military dictatorship, were becoming more Taiwanese. 'It was quite a surprise,' he said. 'I didn't realize that we had judges who were so progressive and willing to bear political responsibility – our case was a highly political case, it wasn't just a pure judicial case.'

Around the time that I met with Lin to interview him for this book, both of the pan-blue candidates running for president, Hou Yu-ih of the KMT and Ko Wen-je of the Taiwan People's Party (TPP), vowed to revive the trade agreement if elected.

'It's like Voldemort coming back again,' Lin said with a laugh, referencing the villain of the Harry Potter books. 'I think this shows that we're standing at the crossroads of Taiwan's development,' he said, referring to the choice of either integrating with China or retaining sovereignty and increasing engagement with the international community. 'We really haven't escaped from these two options.'

Lin had joined the DPP in 2019 over concerns about the local elections of 2018. The KMT's victory in a majority of mayoral and county magistrate elections signalled a potential reversal of the progress made under DPP-led reforms. What concerned Lin most was the meteoric rise of Han Kuo-yu – the former KMT legislator who assaulted Chen Shui-bian in parliament. Han's unexpected ascendancy, especially as the 2020 presidential election approached, brought into sharp focus the possibility of the KMT regaining the presidency. Should that happen, Lin thought, the KMT would roll back all the reforms enacted during the Chen and Tsai presidencies in areas such as transitional justice, same-sex marriage legalization, pension reform, green energy initiatives and the promotion of a more Taiwan-centric education curriculum. These reforms, while championed by the DPP, also deeply resonated with broader civil society during the Ma years.

It's one thing to shout slogans at a protest or occupy the legislature, and something else to work for a political party whose relevance is determined by popular elections, as Lin has discovered. 'My activist friends push me to be more vocal about the issues they care about and to be willing to challenge party decisions,' he said. He also noted that the party establishment 'needs to understand that I'm not coming here to take control of everything or make a revolution within the party. It's very difficult to balance.'

Tsai Ing-Wen and the DPP's Return to Power

Few if any Taiwanese figures have been better at the tricky act of balancing political goals with political realities than Tsai Ing-wen. In 2016, she rode the energy of the Sunflower Movement into the presidential office with a mandate of her own. Tsai pledged to reduce economic reliance on China while expanding relationships with other countries, primarily democracies such as the US and Japan. During her eight years as president, Tsai would have to navigate the challenging geopolitical currents shaped by the administrations of Obama, Trump and Biden in the US, alongside a more threatening China under Xi Jinping.

Most people would not want this job, including Tsai Ing-wen earlier in her life, when she had zero interest in politics. Her father was a native of rural Pingtung in Taiwan's deep south while her mother was born and raised in Taipei. Tsai herself was the youngest of eleven children, all born within a thirteen-year span. While some people might feel lost with ten older siblings, Tsai revelled in the freedom of low expectations.

'There was no pressure, because my older sisters and brothers were all doing very well at school,' she told me when we met in October 2024. 'There was no point in them asking me to prove that my parents have good genes.' Tsai described her early education as 'miserable'. 'I was always behind at school,' she said. 'My parents were not expecting much for me, they just wanted me to have a happy life.' As a young girl, Tsai had an endless supply of hand-me-down books from her elder siblings. That led to her being literate earlier than others her age, which meant she was usually the youngest in her class in her early years. 'I was always having difficulty at school, probably because I started earlier than the other kids,' she told me, almost fondly.

Aside from endless books and low pressure, Tsai's large family helped her learn the art of negotiation. 'When you have so many brothers and sisters to deal with, you need to develop some sort of interpersonal skills,' she told me with a laugh. 'I think that was helpful for my later career.' Perhaps fittingly, Tsai first entered the public eye as a trade negotiator. In the early nineties, President Lee Teng-hui had appointed her as his chief legal counsel for a very important project. She was to negotiate Taiwan's entry into what was originally the General Agreement on Tariffs and Trade, which would become the World Trade Organization (WTO). After more than a decade of negotiations led by Tsai, Taiwan joined the WTO in January 2002 under the awkward name of 'The Separate Customs Territory of Taiwan, Penghu, Kinmen and Matsu'.

Unwieldy monikers aside, Taiwan's entry into the WTO was a significant milestone. The WTO is one of the only major international organizations, along with Asia Pacific Economic Cooperation (APEC) and the International Olympic Committee (IOC), in which Taiwan is a full member. It would not be the last time Tsai would help Taiwan increase its global presence.

'Oh, I enjoyed that part of my life,' Tsai said, smilingly recalling her days as a negotiator in the 1990s. 'I had the opportunity to focus on technical issues and didn't have to think about politics – that was Lee Teng-hui's job.' Tsai preferred the detail-focused work of hammering out trade deals: 'It never occurred to me that I would become that involved in politics,' she told me. However, it was through dealing with different government ministries that she learned their roles and concerns. 'I was actually taught by them how government functions,' she said.

Being a trade negotiator would prepare Tsai for her gradual move to electoral politics years later. 'If allowing greater access to agricultural imports from the US, for example, the farmers here will be affected,' she said, becoming more animated. 'So how are you going to calculate the cost? And how are you going to compensate the local farmers in order to give foreign imports some market share? This was a very important education for a future president.'

Tsai's parents grew up in Taiwan during the Japanese colonial era, and her father went to Manchukuo, Japan's colony in Manchuria, to train as a mechanic. He returned to Taiwan after Tokyo's surrender in 1945, coming home to a country under KMT control. Under such a grim political situation, he focused on developing his auto repair business. One of his competitive advantages was that he was one of the few Taiwanese at that time with extensive experience in repairing foreign automobiles, which was the most lucrative segment of the market.

'Foreigners, or people who were better off, would often seek him for help,' Tsai explained. Her father's initial success would provide the seed capital for venturing into the hotel business, and then property development. 'People like him had no interest in politics at all,' Tsai said. 'They were looking after their families and themselves, just trying to survive during a long period of major changes.' Her mother was a nurse and tailor who was 'relatively more metropolitan than my father, but her family wasn't well-off, so she had to work'.

While her parents grew up under Japanese rule, Tsai was born in 1956 and was spared the worst excesses of the White Terror. 'When I was growing up, the Nationalists were running this place, so I pretty much received a Chinese education,' she told me. Her parents and others of the Japanese-educated generation of Taiwanese who experienced the trauma of 228 'probably had different feelings and perceptions' of KMT rule, she said in her typically understated manner.

When Chen Shui-bian appointed Tsai to head the Mainland Affairs Council in 2000, she would move from trade negotiations to something even more political – cross-strait relations. She took the helm just as Su Chi unveiled his 1992 Consensus fiction. For Chen, the DPP's first president, it would be difficult to have someone from the KMT represent his administration in unofficial talks with China. But Tsai, who did not belong to any party yet and had worked in a non-political role under Lee Teng-hui, emerged as an ideal candidate to lead.

'I seemed to be someone who fit the job description,' Tsai said. 'I had experience with negotiations and had legal training, and had worked with Lee Teng-hui as his National Security Adviser for about a year. Chen Shui-bian thought that having me at the Mainland Affairs Council would bring some sort of continuity.'

Tsai was not entirely unfamiliar with China. In October 1998, she had joined a trip with a government delegation to the country. In Shanghai, she participated in negotiations on cross-strait exchanges at the Peace Hotel on the Bund. During the delegation's visit to Beijing, both sides took a group photo, including Tsai, at the Biyun Temple. Despite growing up in Taiwan, Tsai had been taught more about Chinese culture and history than Taiwanese culture and history. Seeing the place that the KMT had told her was her motherland for the first and only time made a deep impact. 'It was an eye-opening experience for me,' she said, adding that she was impressed by the scale of both Beijing and Shanghai. 'If I were younger then, I would not have minded living there for a while, to get a better feel for Chinese history.'

Unlike when Tsai became president in 2016, Beijing had little leverage in 2000 to pressure the Chen administration to endorse the 1992 Consensus. China was still climbing out of poverty, after all. Most importantly, the PLA had no realistic ability to launch a military invasion on Taiwan. China needed to develop before it could indulge its expansionist dreams.

'Back then both sides felt there was a need to move the relationship forward, and they were more willing to exercise flexibility,' Tsai told me. 'So the key phrase was "One China Principle", and different people have different positions on that, and they just allowed those differences to exist, which started this process.' When Chen Shui-bian won the 2000 election, he 'probably expected a continuation of those talks', Tsai said. But as Chen, and later Tsai and Lai would learn, Beijing has never met a DPP president it didn't despise.

Despite Su Chi's 2006 admission that he fabricated the 1992 Consensus, both the CCP and KMT have used it consistently since 2000 to constrain each DPP president and attack them as troublemakers and even race traitors. 'Rejecting' making Taiwan subordinate to China can then be spun as being defiant or unwilling to compromise. Beijing's refusal to speak with President Tsai or her successor Lai Ching-te unless they endorse the 1992 Consensus is asking the DPP to give up the game before it even starts, which bodes poorly for the possibility of any substantial interactions between Taipei and Beijing for the remainder of Lai's first term, which ends in 2028.

Tsai explained her interpretation of the 1992 Consensus, which guided her presidency: 'Both parties agreed that we should move the relationship forward – that's the consensus. It's pretty much a reflection of what people had in mind back then, and their willingness to be more flexible. And that was the kind of flexibility that we were looking for.' They would not get it.

Tsai's 2016 presidential victory was coupled with the DPP taking control of Legislative Yuan for the first time. In effect, it was the first time since the Dutch arrived in 1624 that Taiwanese-identifying people had full control of a government of Taiwan. I asked Tsai how the historic nature of her and the DPP's victory felt. After taking a deep breath and pausing a moment to revisit that feeling, she exhaled.

'I was thrilled,' she said. 'But there were a lot of expectations. There were so many things that people wanted the government to do.'

I then asked how she thought things had changed in Taiwan during her eight years as head of state: 'Taiwan today seems to be a better-known place,' she said. 'In the past, when I told people I was from Taiwan, they thought I meant Thailand. Today, people know what Taiwan is about.' People are prouder of being Taiwanese now, she added, citing the country's achievements in technology, the economy and democracy.

Despite how far Taiwan has come, there is no doubt that it is currently in a very tight spot. How does Tsai feel about the road ahead for her country? She spoke of national unity in optimistic terms reminiscent of her former presidential adviser, Su Beng: 'The future is always unknown, with a lot of uncertainties, but we have to deal with these unknowns and uncertainties, and get ourselves prepared,' she said. 'The important thing is that Taiwanese people must have faith in themselves. If we can come together to face challenges, we'll be able find ways for Taiwan to continue to survive and thrive.'

If Taiwan is to do so, it will have to resist, one way or another, the growing economic and military might of the Chinese Communist Party.

CHAPTER 9

Taiwan in the Balance

In September 2024, on the eve of the seventy-fifth anniversary of the founding of the People's Republic of China, Xi Jinping – arguably the world's most powerful individual – addressed an audience of more than 3,000 officials, retired cadres and foreign VIPs. They had gathered within the cavernous Great Hall of the People, a hulking Stalinist building on the west side of Tiananmen Square used for official state functions.

Xi stood alone on the massive stage beneath the red-and-gold seal of the PRC. Giant red curtains hung at both sides of the seal, slanting inwards from both sides in a V-shape, leading the guests' eyes to Xi at the centre. At the top of the left curtain was a giant '1949', with '2024' on the right. The PRC was supposed to end dynasties and feudalism in China, yet Xi's demeanour was that of an all-powerful emperor addressing his court. In his speech, he vowed once again to absorb Taiwan into China. 'Taiwan is China's sacred territory,' the dour-faced Xi said in his near-monotone from behind the podium. 'Blood is thicker than water, and people on both sides of the strait are connected by blood.'[1]

Xi's words were cut from the same cloth as the language that Adolf Hitler had used regarding Austria in his political manifesto, *Mein Kampf*: 'German-Austria must be restored to the great

German Motherland,' Hitler wrote. 'People of the same blood should be in the same Reich.'[2]

Calling for increased cross-strait economic and cultural exchanges, Xi said the 'complete unification of the motherland' was 'an irreversible trend, a cause of righteousness and the common aspiration of the people'.

'No one can stop the march of history,' Xi continued, exhorting all to 'resolutely oppose Taiwan independence separatist activities'.

Taiwanese president Lai Ching-te delivered a rebuke to China's claim to be Taiwan's 'motherland' less than a week later, using history to point out inconsistencies in Beijing's rhetoric. 'In terms of age, it is absolutely impossible for the People's Republic of China to be the motherland of the Republic of China's people,' Lai said. He was referencing the fact that the ROC government was founded in 1911, compared with the PRC in 1949.[3] 'One of the most important meanings of these [10 October National Day] celebrations is that we must remember that we are a sovereign and independent country,' he added.

Speaking to Taiwanese media a month earlier, Lai had brought up another historical point. He told reporters that if China's claim on Taiwan was genuinely about territorial integrity, then Beijing should also demand the return of lands taken from the Qing Dynasty by Russia via treaties signed in 1858 and 1860.[4] At that time, Russia threatened the Qing with war if Beijing did not cede what was known as Outer Manchuria, but now is part of the Russian Far East. It was a windfall for Russia, expanding its territory by more than 900,000 square kilometres and giving it the Qing naval base of Haishenwai (now Vladivostok), while also rendering what was left of Manchuria landlocked. Taiwan's total area is just over 36,000 square kilometres, roughly 4 per cent the area of what Russia took from the Qing. If perceived territorial integrity is as important as China claims when discussing Taiwan, why not also retake Outer Manchuria? 'Russia is now at its weakest, right?' Lai asked.

'China's intention to attack and annex Taiwan is not because of what any one person or political party in Taiwan says or does,' Lai said. 'It is not for the sake of territorial integrity that China wants to annex Taiwan.'

Russia, whose invasion of Ukraine has pushed it closer to China, was accused by Taiwanese authorities of launching a cyberattack on the Taiwan Stock Exchange and local financial companies the day after Lai's comments.[5]

While the PRC's founding celebration is on 1 October, the ROC government in Taiwan commemorates its founding nine days later, on 10 October. In his National Day speech, from a large stage in front of the presidential building, Lai also told a crowd of thousands of Taiwan's right to maintain its sovereignty. As he did in his inauguration speech months earlier, he pragmatically embraced Taiwan's ROC government. 'The Republic of China and the People's Republic of China are not subordinate to each other,' he said. 'On this land, democracy and freedom are growing and thriving – the People's Republic of China has no right to represent Taiwan.'[6]

A subsequent Taiwanese poll found that 80 per cent of respondents supported these statements.[7] But China would not react well. Four days after Lai's speech, the PLA abruptly launched a large-scale military drill, 'Joint Sword – 2024B', practising for a blockade of Taiwan. The joint forces exercise included mobilization of the PLA Navy, Air Force, Rocket Force and China Coast Guard.

Among the military assets deployed by China were 153 aircraft and thirty-six naval and coastguard ships. More than half of the aircraft (which included jets, helicopters and drones) entered Taiwan's Air Defence Identification Zone (ADIZ) – a self-declared buffer zone that is outside of official air space. ADIZ incursions by China have increased rapidly since 2019, from fewer than twenty in 2019 to more than 3,000 in 2024.

Of the thirty-six Chinese seaborne vessels, twenty-five sailed to within 24 nautical miles of Taiwan's coast – a distance considered by Taiwan officials as crucial to guarding against attacks in its waters. China's aircraft carrier *Liaoning* lingered further off Taiwan's east coast. Chinese military activity was concentrated in nine areas: near the east coast cities of Hualien and Taitung, off Keelung and Taipei to the north, Taichung and Kaohsiung to the west and south, as well as near the offshore islands of Kinmen, Matsu and Dongyin. It was the fourth time in just over three years that China had conducted such large-scale exercises around Taiwan.

The first drills were launched in August 2022, following the visit of US House Speaker Nancy Pelosi to Taipei. China conducted two more joint exercises following then President Tsai Ing-wen's meeting with US House Speaker Kevin McCarthy on American soil in April 2023, as well as after Lai's May 2024 inauguration. Concerns are rising in Taipei that China's capabilities to turn unannounced drills into a real blockade – itself considered an act of war – or the beginning of an invasion, are growing. 'Although we still believe that war is not imminent and not inevitable, their capacity to switch from exercises to war is really strengthening,' a senior Taiwanese military official told the *Financial Times*.[8]

A Chinese Invasion

No one knows exactly what a Chinese invasion of Taiwan might look like in the beginning, although there are some certainties. Regardless of how many missiles China fires at targets on Taiwan's densely populated west coast, how many bombs it drops or how many drones it sends across the strait, it will not achieve its goal of complete control until it has enough boots on the ground. Estimates of how many soldiers would be needed to subdue Taiwan's twenty-three million people range from around half a million to two million. There is only one plausible way to deliver such numbers, and that is by ferrying them across the Taiwan Strait. It would be the largest attempted amphibious invasion in human history.

For comparison, the current record holder, the Allied invasion of Normandy in 1944, saw more than 150,000 soldiers crossing the English Channel, which is roughly one quarter of the width of the Taiwan Strait. Considering the possibility that the US and Japan could get involved in Taiwan's defence, with potential assistance from the Philippines and Australia, speed would be vital in order to avoid a protracted conflict that could morph into a wider war. Delivering as many personnel and as much equipment as possible in the initial wave would be paramount.

There is no doubt that Taiwan's military is dwarfed by the PLA, but geography is very much on the side of planners in Taipei. The wide Taiwan Strait is famously choppy, and typhoons and other inclement weather are common. Climate change has also rendered the traditional typhoon season of May to October less certain – on 31 October 2024, Typhoon Kong-rey, which had formed quite rapidly in the preceding two days, pounded Taiwan and its surrounding waters. Taiwan also has a limited number of ports that could service incoming Chinese vessels. These include the ports of Keelung, Taipei, Taichung and Kaohsiung – in the instance of an invasion by China, they could be destroyed by the ROC Military to deny access. That would leave amphibious beach landings as the main option. While most of the rugged east coast drops straight down from the mountains, the waters off the west coast are shallow well into the strait, rendering it difficult to get soldiers from larger ships to shore. There are also not many beaches on the west coast where such landings would be feasible, and they are much smaller than Omaha Beach in Normandy, where the bulk of the Allied force was landed.

Chinese planners are aware of this, and China's world-leading shipbuilding industry, which is behind the largest-ever naval build-up during peacetime, is rapidly expanding the PLA Navy's amphibious fleet. As of 2024, the PLA Navy had three active Type 075 amphibious assault ships, with another still being fitted out. Four more are planned, and they are being built at the rate of one every six months. These large vessels resemble aircraft carriers, with a top deck that can be used by attack helicopters and a well deck within the hull that can discharge amphibious tanks or hovercraft.[9] The first Type 076 vessel, an upgrade from the Type 075, was under construction at the time of writing this book. Significantly larger than its predecessor – nearly the size of three American football pitches – the Type 076 will feature new capabilities. Most noteworthy is the catapult for launching fixed-wing aircraft, something no other amphibious assault ships currently have. China's third and newest aircraft carrier, the *Fujian*, was the first to feature a catapult.[10]

While these vessels could inflict substantial damage, they don't solve the problem of moving large numbers of troops across the strait. China, which boasts the world's largest civilian ferries, is also working on that. In 2021, it was discovered that some of its biggest ferries, typically plying routes between northern China and South Korea, had been participating in amphibious invasion drills with the Chinese military on beaches in southern China. These colossal vessels, which are more than twice the size of their largest American counterparts, had been retrofitted with ramps that enable amphibious vehicles to roll on and off at sea. Chinese planners have previously said they have more than sixty ferries that could serve such a purpose.[11] As civilian vessels incapable of defending themselves, however, they would be vulnerable to attack.

'If they use these types of ferries in the first wave, the risk is very high,' Shen Ming-shih, a research fellow at the Institute for National Defense and Security Research (INDSR) in Taipei told me. 'These kinds of ships would be sitting ducks.' Within three to five kilometres from Taiwan's shores, the ferries would be at the mercy of Taiwanese artillery and anti-ship systems, Shen said.

China would likely need to destroy more than half of Taiwan's anti-ship systems before deciding to deploy civilian ferries filled with troops and amphibious tanks, he added.

While the PLA is upgrading its hardware and capabilities, the fact remains that its last successful invasion was of Beijing itself in 1989, in response to the Tiananmen protests. Its last invasion of another country was in 1978, when it poured into northern Vietnam via their shared land border. China lost that war.

The early missteps of the Russian forces in Ukraine highlighted the gap between perceptions and reality of the Russian military. While China may seem much stronger, it doesn't have Russia's battle experience. Joint exercises such as those used to intimidate Taiwan are therefore valuable in familiarizing different branches of the PLA with working with each other. Many observers believe that Xi Jinping would prefer the Chinese military to have more experience in joint operations before attacking Taiwan, but, as with many things involving Xi and his black-box government in Beijing, this is informed speculation, but speculation nonetheless.

As the armed wing of the CCP, the PLA's primary task above all others – even invading Taiwan – is keeping the party in power. This requires maintaining control over more than 1.3 billion Chinese, many of whom are increasingly disgruntled for both political reasons as well as economic ones. As a result, it needs the buy-in of the Chinese people. After eight decades of CCP propaganda declaring that Taiwan is an inseparable part of China, most Chinese citizens appear to believe that Taiwan is Chinese territory. That said, China does not permit independent political surveying and it would be risky for an individual to say they didn't think Taiwan was China or even that it should be its own country – something I've heard many Chinese people say in private conversations. As a result, it is impossible to estimate how many Chinese would genuinely support a war over Taiwan's territory that would also likely devastate the economy and kill large numbers of Chinese soldiers.

In its war preparations, Beijing must therefore also consider domestic public opinion, which is much more informed and educated than during the 1978 invasion of Vietnam. So how is Beijing marketing a potentially catastrophic war to its citizens?

In 2021, the Chinese scholar and government adviser Jin Canrong gave a speech to a packed auditorium in Beijing on the ten biggest benefits of a war to take Taiwan. Jin's speech is not necessarily official government thinking, but the fact that it occurred in Xi Jinping's highly censored China suggests that the government doesn't disagree with his main points. It is a poignant snapshot of the things that Chinese elites are being told about why a war for Taiwan would be good.

Jin is a professor and associate dean at the School of International Relations at Renmin University in Beijing and is a well-known political commentator in China. He is considered one of the country's foremost experts on the US and, his nationalist opinions aside, is also known for his claim that Aristotle never existed. Seated on a stage in a white office shirt at a desk with his laptop in front of him, Jin made his case for invasion. The mostly male audience, all in masks, listened attentively.

The first benefit of invading, he said, would be breaking through the First Island Chain. 'After we conquer Taiwan, we'll be directly facing the Pacific Ocean, right?' he asked. 'The entire geopolitical position will be much better.'

Second, taking Taiwan, he said, would end China's 'century of humiliation' at the hands of foreign powers that began with the Opium War of the mid-nineteenth century. He neglected to mention the much larger chunks of territory that Russia took from China in 1858 and 1860. 'Us Chinese will have a much healthier and more confident mentality, which would be very good for our future development,' he claimed.

Third, conquest of Taiwan would firmly establish Xi Jinping's 'new era of socialism with Chinese characteristics'. This is a necessary mention of Xi, the most powerful Chinese ruler since Mao. By taking Taiwan, Xi would be elevated above Mao to the top of the CCP pantheon.

The fourth benefit of invading Taiwan, he said, is that China's military would rise in status. Jin notes that in PRC history, all the battles the PLA has fought have been in 'self-defence' or 'counter-attack' on China's doorstep, glossing over its unprovoked invasion of Vietnam. 'Taiwan will be the first case of reclaiming territory my grandfather lost, no?' he asked. After the Korean War, he continued, nobody dared invade China, while Taiwan would be a war of unification and rejuvenation. With a sinister smile, he said that there would be personal benefits for PLA soldiers and officers, claiming, 'It will be much easier for them to find a wife.' Soldiers' salaries will increase severalfold, he also asserted.

The fifth big benefit seemed to be of more interest to the audience: Taiwan's world champion chipmaker, TSMC. 'After taking back Taiwan, we can immediately nationalize TSMC,' he declared, at which the audience laughed appreciatively. Jin's statement, however, ignored the fact that TSMC and its supply chain would all but certainly be destroyed in an invasion scenario, whether by China or even Taiwan or the US seeking to prevent China from seizing the world's most important company. Additionally, TSMC is dependent upon a real-time supply chain of highly specialized chemicals and materials from the US, Japan and Europe, which any war would sever. TSMC chairman Mark Liu has said that an invasion would render TSMC inoperable.[12]

A sixth benefit, Jin asserted, would be improved political stability for China. 'I personally believe Taiwan is the main source of all the "colour revolutions" in mainland China,' he said, referring to social disruptions of recent years. 'Of course, [Taiwan is] working with the US,' Jin elaborated, going on to accuse Taiwan of funding or supporting the Falun Gong spiritual movement that has been banned and persecuted by the CCP; Chinese democracy activists;

as well as independence movements in Hong Kong, Xinjiang and Tibet. Taiwan is also spreading misinformation on the Chinese internet, he says. 'They also attack influencers like me who are spreading positive energy,' he said, pointing at himself. 'Taiwan regularly attacks me for nothing.'

Benefit number seven was improved social stability. Jin accused Taiwan of being behind telecoms fraud, which is rampant in China. Jin said '90 per cent' of telecoms fraud in China is done by Taiwanese. If this was true, it would be an impressive feat for twenty-three million people to do that to the 1.3 billion people next door. He went on to make the nonsensical claim that the Taiwanese brag about having three big non-polluting industries: tourism, pornography and telecoms fraud.

An improved coastal economy was benefit number eight. Describing solving the 'Taiwan problem' as the PLA's number-one task, he complained that the precious resources in coastal provinces that were allocated to military preparations to conquer Taiwan could go elsewhere. Maintaining China's numerous coastal military bases has affected economic development, he said, and after taking Taiwan, those bases could be moved to Taiwan's 'backward' east coast. Jin claimed that the frequent civilian flight delays experienced by eastern and southern Chinese cities were Taiwan's fault, accusing Taiwan of harassing China with its fighter jets. (It is worth noting that during the Cold War, ROC fighters based in Taiwan did harass China, but that was prior to democratization.)

'Whenever Taiwan, Japan, the US and Australian aircraft move around, our jets have to immediately take off, which affects civil aviation,' Jin said. 'So after the Taiwan problem is solved, our country's flights will become much more punctual,' he said to laughter and applause. 'Then our aircraft can fly toward Japan and Australia for exercises and delay their flights.'

The ninth benefit of conquering Taiwan would be the lightening of a major diplomatic burden for Beijing, Jin said. Not only do the

US and other countries 'bully' China over Taiwan, but developing countries, specifically from Africa, also take advantage of China's chequebook diplomacy, he claimed. Speaking of African leaders who visit China, he said, 'They've got nothing to do, just having a look around Beijing, eating nice food and then when meeting with Chairman Xi, they have only one thing to say: "We resolutely support the one-China policy, Taiwan is an inseparable part of China."' After they've said that, he said, they immediately say, '"Boss, give us money."' He and the crowd laugh. 'They really do, they have no shame,' he said. Subsuming Taiwan would end China's need to bribe world leaders to support Beijing's territorial claim.

Saving the biggest benefit for last, Jin said a successful invasion would change the international landscape overnight: 'If we determinedly take Taiwan, with the US unable to do anything but yell uselessly – our international image will have changed,' he said. 'Everyone will know the original boss is no good any more, he's grown weak – the new boss is here.

'I believe after we take Taiwan, the vast majority of countries will begin to turn to China. They don't have principles, they're very snobbish, very practical.'

On the other side of the coin, it's difficult to see many benefits for Taiwan if it were to be invaded and assimilated by China. Taiwan currently has 2.5 times the per capita GDP of China. It boasts a highly educated society with universal health care, unfettered internet, Asia's freest press, same-sex marriage and many other benefits that China doesn't have. The majority of Taiwanese people have no desire to be part of the PRC, but would also like to have a friendly or at least non-antagonistic relationship with their neighbour.

'Since Ancient Times'

More than 1,700 kilometres north of Taiwan, Tiananmen Square is an awe-inspiring symbol of the CCP's power. Originally built by the Qing Dynasty in the mid-1600s, the CCP expanded its area fourfold in the 1950s. The square sits on the south of the Forbidden City, a walled-off city within Beijing where Ming and then Qing emperors ruled China from 1420 to 1924. Following the vanquishing of the KMT's ROC party-state in 1949, the CCP built the obelisk known as the Monument to the People's Heroes to the square's south, lionizing the victors of the Chinese Civil War. Tiananmen Square is where hundreds of thousands of Chinese gathered to hear Mao declare the founding of the PRC in 1949, as well as where as many as a million Chinese mourned his passing in 1976. The Great Hall of the People sits on the square's western edge, where it hosts rubber-stamp legislative meetings and holds important state ceremonies, often for VIP foreign guests. Within the hall, each administrative region of the PRC has its own pavilion – centrally administered cities like Beijing or Shanghai, provinces like Sichuan or Guangdong, or autonomous regions like Tibet or Xinjiang. All of the pavilions depict places administered by the PRC, save for one: the Taiwan Pavilion. At the cavernous room's entrance, a gold-framed red stone slab greets guests with the Taiwan story, told in chiselled gold Chinese characters. Fifteen large ideograms on the right announce all the reader needs to know: 'TAIWAN HAS BEEN CHINA'S SACRED TERRITORY SINCE ANCIENT TIMES'.

As confident as it is vague and misleading, it is a memorable mantra that China's billion-plus population have heard in one form or another throughout their lives. It is also the message that the CCP and the KMT have propagated to the world since the early 1940s. No matter how dubious the history is, when you have two highly influential organizations repeating the claim for over eighty years, it will naturally take root.

The CCP continues to prepare audiences both domestic and international for what they are supposed to view as a historic inevitability: the 'return' of Taiwan to China, by any means necessary. In late 2021, platforms affiliated with the Chinese government promoted a video for the nationalist song 'Going to Taiwan in 2035'.[13] Bass-heavy traditional Chinese drums rumble as a montage of popular tourist destinations in Taiwan changes along with the beat. A Chinese flute joins in and suddenly a child and adult walk hand-in-hand in a nondescript park. A deep, manly sounding voice sings, inviting the listener to meet him in Taiwan in 2035, as an image of a Chinese high-speed train suggests that such a trip from China will be part of the national rail network by then. Taiwan's Alishan, Sun Moon Lake and the islands of Penghu are all mentioned, but no Taiwanese cities are name-checked in the song's lyrics. The only city mentioned is Beijing. Throughout the video, we see only the backs of Chinese people as they frolic in a version of Taiwan that was fabricated in a Chinese studio. The most disturbing aspect of the video, aside from the implied end of Taiwan's sovereignty, is that not a single Taiwanese person is depicted during its four minutes. The message is as clear as it is chilling: Taiwan is ours and Taiwanese people do not exist.

And what if China were to win, realizing Xi Jinping's dream of national rejuvenation? For both Taiwan and the world at large, a successful invasion would be catastrophic. Given its campaigns of cultural genocide against Xinjiang and Tibet, and ongoing political repression of Hong Kong, China is familiar with the tools for subjugating unwilling populations. In 2022, two of its top diplomats, the Ambassador to France Lu Shaye and Ambassador to Australia Xiao Qian, both made comments asserting that China intends to 're-educate' Taiwan after it 'liberates' its people.

'After reunification, we are going to do re-education,' Lu told French TV station BFMTV shortly after then-House Speaker Nancy Pelosi visited Taiwan.[14] Days later, Xiao, who was participating in an event with the National Press Club of Australia, was asked about Lu's comments: 'I haven't read such about such an official policy. I think my personal understanding is that once Taiwan is united, come back to the motherland, there might be process for the people in Taiwan to have a correct understanding of China,' the *Australian Financial Review* reported Xiao saying.[15]

Xiao's comments echoed Chinese government rhetoric on the concentration camps in which a million or more Uyghurs have undergone so-called 're-education' since 2016.[16] The US and other governments around the globe have defined Beijing's policy against the Uyghurs as genocide. Just as China wants Xinjiang but doesn't regard Uyghurs as full citizens, its claim on Taiwan also appears to be solely territorial. The Chinese do not want to integrate Taiwanese people, who are often viewed as corrupted Chinese, just as the KMT viewed them upon its arrival in 1945. A popular hashtag for Chinese nationalists when discussing Taiwan online is #留岛不留人, or 'keep the island but don't keep the people'.

It is the will of the Taiwanese people, after all, that stands in the way of all the benefits of a country that the PRC has never controlled and to which it has no legal claim. But it is within Taiwan itself that the CCP has one of its greatest assets – its old nemesis, the Chinese Nationalist Party, or KMT.

A Tale of Two Chinese Ethnonationalist Parties

When Xi Jinping and Ma Ying-jeou met at the Shangri-la Hotel in Singapore in 2015, the two heads of state did so in their roles as heads of their parties, the CCP and KMT. The two men entered a packed ballroom from opposite sides. Ma's tie was blue, the colour associated with the KMT, while Xi's tie was CCP red. They met in the middle of the room, where they shook hands for a full minute, barely making eye contact as they smiled for the flashing cameras. Xi and Ma then gave statements to the press before holding a closed-door meeting.[17]

Reading from a prepared statement, Xi stressed the CCP's blood-and-soil view that, because most Taiwanese people are of Chinese descent, then Taiwan should belong to China, regardless of the Taiwanese people's feelings on the matter. 'We are brothers who are still connected by our flesh even if our bones are broken,' Xi said of China and Taiwan. 'We are a family in which blood is thicker than water.' In an apparent reference to the US, Xi warned that 'no force can pull us apart', as Ma looked on in seeming approval.

For his part, Ma put forth five ways of maintaining peace between the two sides, with the first and most important being 'consolidation of the 1992 Consensus', a roundabout way of rejecting any possibility of a Taiwanese state.[18] Xi listened with a contented look on his face.

International media reports breathlessly framed the summit as the first time that the leaders of China and Taiwan had met since 1945. Ma's approval ratings were near single-digit levels in Taiwan and the meeting – just weeks before the 2016 presidential and legislative elections – was an apparent attempt to boost the fortunes of the KMT, which had been trounced in local elections the year before.

Nothing new was announced after their meeting. Both men reiterated their parties' positions that Taiwan and China are part of 'one China' and that Taiwanese self-determination was impermissible. They then split the bill on a relatively humble meal of spicy noodles, fried asparagus and crayfish, all washed down with sorghum liquors distilled in their respective regions of their respective imagined Chinas.

While much of the discourse about the meeting focused on the Taiwan-and-China-are-making-friends angle, seasoned observers of Taiwan and China saw it as further rapprochement between the two Chinese ethnonationalist political parties on either side of the strait. There was no denying it: with its popularity flagging in Taiwan, the KMT was moving deeper into the orbit of the CCP to stay politically relevant. Nevertheless, it was a historic occasion, given that it was the first meeting between the chairmen of the two parties in seventy years.

The last time that the leaders of the CCP and KMT had met was in August 1945, when Mao Zedong and Chiang Kai-shek spent seven weeks discussing the future of China. They met in Chongqing, known then as Chungking, where Chiang had established his wartime capital after abandoning Nanjing to the savagery of the Japanese Imperial Army.[19] Chiang was not happy that the Truman administration's ambassador to China, Patrick J. Hurley, had used an American plane to escort Mao to Chongqing from the Communists' stronghold in northern China. But Chiang had little choice and he knew it: the war was over, so he could no longer threaten to sue for peace with Japan. His leverage over Washington was gone.

The US hoped to see some form of power-sharing agreement that would avoid a rekindling of the civil war that had been interrupted by Japan's invasion of China. By this time, the State Department had allowed itself to be duped by Mao's claims that the Communists were in fact democratic. And it had grown tired of Chiang.

Mao and Chiang attempted to project an image of unity, at one point rising to clink small glasses of sorghum liquor as they toasted Japan's surrender just days earlier. Neither man was interested in sharing power, however, and the following year would see the return of full-scale war to an already devastated China.

Unlike 1945, the US did not broker the Xi–Ma summit of 2015. The KMT and CCP had normalized relations a decade earlier after it had become clear that the genie of Taiwanese identity had been let out of the lamp, threatening both Chinese political parties. Despite the previously bloody rivalry between the KMT and the CCP, they have shared roots and were alarmed by the rise of Taiwanese identity in the democratic era. The erstwhile enemies had more to gain by working together than by letting Taiwan's democracy continue to develop unchecked.

Roughly a year later, Tsai's congratulatory phone call to the American president-elect, Donald Trump, put the DPP and the US government on a path towards their warmest relations since the DPP's founding. While there is no evidence to suggest that Trump himself cared much for Taiwan beyond its use as a bargaining chip with Beijing – upon taking office he glibly hinted at the possibility that the US might recognize Taiwan – he surrounded himself with China specialists who were very interested in supporting Taiwan, many of whom had deep personal connections to the island nation. Between 2008 and 2016, when Ma and the KMT controlled Taiwan's executive and legislative branches, the Obama administration was more focused on keeping its political and economic relationship with China on an even keel, choosing to ignore Xi's militarization of the South China Sea. That was shaken up after Trump launched his trade and tech wars with China, catching Beijing off-guard with his follow-through on his promise to tariff Chinese goods.

Tsai's re-election and Biden's advancement of Trump's support for Taiwan only further entrenched a new political reality in Taiwan: its two largest political parties have become extensions of, if not proxies for, the larger power struggle between Washington and Beijing.

Roughly half of Taiwan's voters are unaffiliated with any party, creating a massive pool of votes that are up for grabs each election. These swing votes have recently punished incumbent presidents during the midterm elections, in which the magistrates of Taiwan's counties and mayors of its largest cities are chosen. In Taiwan, as in other democracies, midterm elections tend to serve as a referendum on the executive branch or incumbent government's performance since the previous general election.

Tsai and the DPP were handed a strong rebuke in late 2018, with the KMT winning the majority of available seats. On the night of the 2018 midterms, Tsai's party went from controlling thirteen of Taiwan's twenty-two counties and municipalities to controlling only six. She appeared stunned by the results, but took responsibility for the battering and stepped down as party chair on live television. It was her first progress report as president, and Taiwan's electorate had not given her a passing grade.

After Tsai's sombre departure from her press conference,

media attention in Taiwan shifted to the island's south. Han Kuo-yu had pulled off a massive upset to become Mayor of Kaohsiung, an office the DPP had dominated for the previous twenty years. Often stumping in blue dress shirts with rolled-up sleeves, Han had campaigned as a straight-talking man of the people. His previous job as general manager of the country's largest wholesale produce market helped him burnish this image, although he had been a legislator from 1993 through 2002. Despite not being taken seriously within the KMT, Han managed to tap into the psyche of many Taiwanese in 2018, emerging from the political wilderness to become a juggernaut who led the party to an unexpected victory deep in enemy territory. Several KMT candidates that he had endorsed in other races rode the 'Han wave' into office. His blunt and often politically incorrect swagger found its moment in Taiwan as populism was making inroads in other democracies. Some of his supporters donned caps emblazoned with 'MAKE KAOHSIUNG GREAT AGAIN'.

The night of his mayoral victory, a crowd of tens of thousands thronged near the stage where he would make his acceptance speech. The excitement upon his arrival was beyond electric. Between the sheer number of people in the streets to support Han and wall-to-wall media coverage, the moment felt like less of a mayoral acceptance speech than it did the launch of his presidential ambitions.

Han had deftly seized upon growing resentment in Kaohsiung which, in the 1990s, was Taiwan's most important city after Taipei and was the world's third-busiest container port. Since then, however, a large-scale shift of trade and investment across the strait to China had hollowed out the city's economy. Lamenting that Kaohsiung had become 'too old and too poor', Han's combination of plaintive sympathy and fiery rhetoric struck a nerve with voters who typically wouldn't vote for a KMT candidate, yet felt the DPP was taking the city's loyalty for granted.[20] Upon completing their studies, local university graduates overwhelmingly chose to head north to other cities such as Taipei, the central Taiwanese metropolis of Taichung or the tech hub of Hsinchu. There simply weren't opportunities in Kaohsiung any more – the city had been left behind. For Han, there was an easy cure for Kaohsiung's ills: China.

'If goods flow out and people flow in, Kaohsiung will become wealthy' was Han's simple yet effective campaign slogan.[21] The goods flowing outwards would be exports to China, and the people flowing into Kaohsiung would be Chinese investors and tourists. Han spoke of building a giant Ferris wheel, a horse racing track and even a Disneyland resort – even though Disney was already partnered with the government of Shanghai in a massive park in the Chinese city. More bizarrely, he vowed to bring Arnold Schwarzenegger to Kaohsiung. Above all else, he said, he would turn Kaohsiung into Taiwan's richest metropolis.

None of this would happen, but Han did follow through on his promise to get closer to China. He set off for China in March 2019, two months after taking office, ostensibly to sell produce and seafood from Kaohsiung to the massive Chinese market in his capacity as mayor. Appearing energized by the star's welcome he received from Chinese media and officials, Han declared his allegiance to the idea that China and Taiwan belong to the same country. He met with the top Communist Party officials in the finance hub of Hong Kong, the casino mecca of Macau and the tech hotbed of Shenzhen. While in Shenzhen, Han held a surprise meeting with Liu Jieyi, the head of Beijing's Taiwan Affairs Office, the department that oversees Taiwan policy. Although claiming that his trip to China was not made for political reasons, Han was having a friendly meeting with the Communist official tasked with bringing Taiwan under Chinese control. As photographers snapped away, Han and Liu sat in chairs on opposite sides of a large floral arrangement. Between their body language and the statements they made before reporters, it was apparent that they did not consider themselves equals – Liu was holding court.

'The two sides of the strait should do everything they can as soon as possible to create a common market and share in the opportunities of development of the mainland of the motherland,' Liu said with a thick Beijing accent and the stilted oration of Communist Party officials.[22]

Before the cameras, Han reiterated his support for the 1992 Consensus. 'I'm not going to lie to you,' Han said in a deferential tone. 'I have always upheld the 1992 Consensus, this is extremely important.' Liu nodded in approval.[23]

Back in Taiwan, the optics of the meeting were initially helpful for Han, who took on the air of a cross-strait statesman. Pro-China media in Taiwan was exhaustive in its coverage of Han's China visit, following his every move. These outlets portrayed him as a pair of steady hands that could manage the strained cross-strait relationship – which they blamed on Tsai, despite her repeated calls for dialogue. Since Tsai took office in 2016, China had cut off communication with her administration, seeking to sideline her for not accepting the 1992 Consensus in her inaugural speech. China had rejected Tsai's overtures due to her insistence that any talks be conducted as equals, a non-starter for Beijing.

Under the administration of Ma Ying-jeou, pro-unification Taiwanese businessman Tsai Eng-Meng, who had made his fortune in China, bought up several prominent print, online and broadcast platforms, creating the Want Want China Times Media Group conglomerate. Tsai Eng-Meng is one of the richest men in Taiwan, and not unlike Rupert Murdoch in the West, has leveraged his media holdings to further his own political agenda, at the heart of which is cross-strait unification. His media outlets often play up Taiwan's Chineseness while erasing Taiwanese identity or portraying it as dangerous.

Despite the power they hold, Tsai and the Want Want China

Times Group have their work cut out for them. Taiwanese identity has become increasingly entrenched since the KMT ended martial law in 1987. Taiwan's National Chengchi University – itself an institution that arrived with the KMT in the 1940s – has conducted surveys regarding identity in Taiwan since 1992. In the summer of 2024, it published its latest results, highlighting that most Taiwanese consider themselves . . . Taiwanese. Indeed, 63 per cent of respondents indicated that they thought of themselves as exclusively Taiwanese. Another 31 per cent said they thought of themselves as both Taiwanese and Chinese, while those identifying as exclusively Chinese were a mere 2.4 per cent. Compare that with the late 1990s, when Taiwanese identity was only 24 per cent, while dual identity was over 49 per cent and Chinese identity exceeded 26 per cent.[24] The further Taiwan gets from the martial law era, the more Taiwanese it becomes.

Despite these odds, in 2018, the pro-China media in Taiwan had two things working in its favour as it helped lift Han Kuo-yu and the KMT. The first is the aforementioned large pool of voters who are not affiliated with any party. This means that people often vote *against* candidates or parties as much as they may vote *for* certain candidates or parties. Notoriously demanding, Taiwanese voters are unafraid to discipline candidates from a party they prefer, even if it means abstaining or voting for a party for which they don't normally have an affinity. The second factor is China's giant economy. Taiwan's economy has been substantially intertwined with China's since the early 1990s, and potentially shifting away from manufacturing goods in China or selling goods to China felt like bad economics for many Taiwanese, although that has changed since.

During both the 2018 local elections and 2020 national elections, pro-China Taiwanese media portrayed Han and the KMT as the preferred alternative to Tsai and her party, which they painted as out-of-touch elitists. These outlets also pushed a narrative that boiled down to: if Han and the KMT win, the China threat will go away and everyone is going to make more money.

Initially, that narrative appeared to work with many Taiwanese. One poll published in May 2019 showed Tsai ahead of Han by a vulnerable 7 per cent margin.[25] But the election was still eight months away, a very long time in Taiwanese politics. As the election campaigns progressed, Han's pro-China stance became more of a liability than an asset. Meanwhile, Tsai's willingness to stand up for Taiwan's sovereignty in the face of pressure from Beijing helped her shake off the setback of the midterm elections.

On 1 January 2019, just weeks before Han's cross-strait trip in March, Xi Jinping declared in his first major policy speech regarding Taiwan that the country and its twenty-three million people 'must and will be' unified with China, even if it took military invasion to achieve.[26] Should Taiwan be willing to renounce its sovereignty and be absorbed peacefully by its neighbour, it could enter into a 'one country, two systems' arrangement similar to that of Hong Kong. When the UK handed Hong Kong over to Chinese rule in 1997, Beijing had promised the former British colony a high degree of autonomy over its legal, political and economic system for fifty years, until 2047. The Chinese government, however, would be in charge of matters of diplomacy and national security.

Not long after Han visited China, the CCP's promise to Hong Kong, already under strain, began to unravel. A proposed bill that would allow the extradition of people from Hong Kong to China's mainland sparked massive protests, including a march of one million people in the territory of seven million. Two months later, two million Hongkongers took to the streets in an even more breathtaking demonstration of their determined opposition. The bill was eventually withdrawn but the damage was already done, with police cracking down on gatherings with increasing violence and brutality, even shooting demonstrators and injuring journalists. Protestors grew increasingly violent in their pushback as well, using Molotov cocktails, three-person slingshots and even bows and arrows to fight back.

Initially slow to criticize the Hong Kong police, Han eventually had to publicly reject the notion of Taiwan entering a Hong Kong-style 'one country, two systems' deal with China. But it was too late for him to shed his pro-China image. Tsai, on the other hand, had responded to Xi's speech at the beginning of the year, just after he delivered it, rejecting it outright: 'It is impossible for me or, in my view, any responsible politician in Taiwan to accept President Xi Jinping's recent remarks without betraying the trust and the will of the people of Taiwan,' Tsai said at a January briefing for foreign journalists.[27] That was the beginning of her comeback and the start of Han's troubles.

When Han formally declared his candidacy for president in the summer of 2019, only seven months after taking office as mayor, many of the Kaohsiung voters who had swept him into office soured on him. The deteriorating situation in Hong Kong hurt Han and boosted Tsai as young people around Taiwan embraced the slogan 'Hong Kong today, Taiwan tomorrow'. Han became more erratic and, echoing Donald Trump's campaign threats against his 2016 opponent, Hillary Clinton, he insinuated that Tsai's domestic infrastructure initiatives were tainted by corruption, and vowed to investigate her if elected president. Not long after Han was confirmed as the KMT candidate to challenge the incumbent Tsai, fellow party members publicly accused him of having issues with drinking, playing mah-jong and philandering. Han has denied these allegations. His campaign rallies became increasingly negative as he fell further behind Tsai in the polls.

In the end, Tsai won by a landslide, receiving more votes than any presidential candidate ever had in Taiwan's short democratic history. High youth turnout skewed in her favour, turning what would have been a convincing victory into a rout. The DPP also retained control of the Legislative Yuan, all but ensuring that Tsai would be able to push her agenda forward for another four years, until 2024. The KMT was down, but not out. Despite losing by a large margin, Han still received 5.5 million votes, a significant increase from the 3.8 million votes cast for KMT candidate Eric Chu in his 2016 loss to Tsai. Another aspect of the election, the party vote, provided solace to the KMT.

In every general election, Taiwanese cast three votes: one for the presidential ticket, one for their local legislator seat and one for their preferred party. The third ballot, known as the party vote, goes towards filling thirty-four of the 113 legislative seats up for grabs, via proportional representation. So, if one party secures half of the party votes, they will receive seventeen of the thirty-four seats allotted. While there are usually only two or three parties that are large enough and organized enough to field presidential candidates, there are many smaller parties that may have a message that resonates with enough of the electorate to justify their inclusion in the legislature.

For the party vote, parties submit a list of at-large candidates who are not running for specific local seats – these are often people who might not be able to win a competitive race on their own. If a party wins one seat via the party vote, the candidate at the top of the list gets the seat. If they win ten seats, the top ten candidates on their list get seats.

In the 2020 election, the KMT received 33.3 per cent of the party vote, only slightly less than the DPP's 33.9 per cent. Among the KMT legislators who were awarded seats via the party list was Wu Sz-huai, a highly unpopular former general who would never have been able to win a contested local legislator seat. Wu was a source of concern for many Taiwanese voters who saw him as an agent of China, and not without reason. In 2016, he attended a speech by Xi Jinping at the Great Hall of the People in Beijing, where, according to Taiwanese media, he sang the national anthem of the PRC. Wu had also commented on Chinese television, offering advice on how the PLA could defeat the US in a military clash over Taiwan. Wu's seat in the legislature granted him access to classified military intelligence, which was a major source of consternation for non-KMT legislators.

The same party list that allowed Wu Sz-huai to be indirectly voted into the Legislative Yuan in 2020 helped get Han Kuo-yu back into action in the next general election in 2024. The KMT's inclusion of Han at the top of their party list not only returned him to the legislature but also set him up to become Speaker of the Legislative Yuan.

Before Han took the gavel, former president Ma Ying-jeou made a trip to China, where he met with Xi Jinping for the first time since their 2015 meeting. Later, a group of seventeen KMT legislators also visited China. Following their return, only one KMT legislator attended the May 2024 inauguration of Lai Ching-te and Hsiao Bi-khim. There is little guessing as to the KMT's intentions. At the time of writing, the KMT-led majority coalition in the legislature was attempting to expand its powers at the expense of the executive and judicial branches, while also moving to freeze or cut portions of budgets of all ministries – including funds for defence and diplomacy. Senior legislator Weng Hsiao-ling, one of the leaders of this push, said in an interview with *Nikkei Asia* of Taiwanese people: 'We are Chinese.' She used the term *zhongguoren* – 'Chinese nationals' – rather than *huaren* – 'ethnic Chinese', precipitating an angry maelstrom on Taiwanese social media.[28] 'The peaceful unification of this country is of course the ultimate goal,' she added.

For anyone steeped in Taiwan's fractious politics, the message was clear – the gloves were off, and the next four years would be a constant battle between a KMT – with a very friendly CCP standing behind it – and a DPP that could no longer pass legislation at will.

CHAPTER 10

Scrubbing Taiwan's Sovereignty

Midday sunlight dappled the East River as I walked towards 42nd Street in Manhattan's Midtown East neighbourhood. It was late October 2019 – just months before the city would be ravaged by Covid-19. My friend Jenny was waiting for me in front of the UN Headquarters, where flags flapped in the gusty wind in front of the General Assembly Building. The imposing glass curtain of its thirty-nine storeys reflected the Long Island skyline across the river.

It was my first visit to the UN, which emerged from the embers of the Second World War with the promise of peaceful resolutions of disputes between states, and self-determination for all peoples. Our guide, a dark-haired man in his thirties with a closely trimmed beard, told our group what the UN sees when it looks in the mirror: a place where every country has a voice, a champion of sovereignty and dignity of all nations, regardless of size. When our guide turned to the subject of the Security Council, he naturally mentioned the five permanent members – the US, Russia, France, the UK and China. 'The winners of the Second World War, more or less,' he said. Jenny and I exchanged glances. Others in the group appeared not to notice the heavy lifting the turn of phrase 'more or less' was doing. The China of today, the authoritarian PRC party-state, was not one of the victors of the war. It wasn't even founded until 1949, four years after the war ended, when Mao's revolution overthrew the ROC government.

The ROC is still technically the government of Taiwan. It is now unrecognizable compared to when it arrived in 1945, having democratized under pressure from the Taiwanese people at the end of the last century. It is now considered Asia's freest democracy – a claim that was bolstered by its eighth free and fair presidential election in 2024. The KMT is now a minority party. Chiang's dream of using Taiwan as a base to retake China is but a distant memory. Most Taiwanese consider themselves citizens of a country they colloquially call Taiwan.

Just months before it seized Taiwan from a defeated Japan, the ROC became one of the founding signatories of the UN Charter. Following Chiang's escape to Taiwan, his regime retained its seat (and veto) on the UN Security Council until his position became untenable. That took twenty-five years, during which Mao's PRC derided the exclusion of the world's most populous country as undermining the UN's legitimacy. It was a valid point – and one that Chiang wilfully ignored, to the long-term detriment of the Taiwanese people, whom he clearly cared little about.

In the 1960s, Chiang's enablers in Washington had encouraged a two-state settlement, with Beijing taking over the China seat and Taipei receiving a seat under the name Taiwan. Chiang refused to the bitter end, spitefully choosing to leave the UN rather than endure the humiliation of negotiating for a lesser seat at the same table as the PRC.

On 25 October 1971, the CCP delegation celebrated with uncharacteristically big smiles as the UN General Assembly passed Resolution 2758, proposed by Albania, a fellow communist state. A large number of delegates from countries around the world whooped and hollered in celebration as the American-assisted campaign to keep Nationalist China in the UN finally failed. The full language of the resolution is as follows:

THE GENERAL ASSEMBLY,

    Recalling *the principles of the Charter of the United Nations,*

    Considering *the restoration of the lawful rights of the People's Republic of China is essential both for the protection of the Charter of the United Nations and for the cause that the United Nations must serve under the Charter,*

    Recognizing *that the representatives of the Government of the People's Republic of China are the only lawful representatives of China to the United Nations and that the People's Republic of China is one of the five permanent members of the Security Council,*

    Decides *to restore all its rights to the People's Republic of China and to recognize the representatives of its Government as the only legitimate representatives of China to the United Nations, and to expel forthwith the representatives of Chiang Kai-shek from the place which they unlawfully occupy at the United Nations and in all the organizations related to it.*[1]

Note that there is no mention of Taiwan or the Taiwanese people in the resolution. Additionally, the 'restoration of the lawful rights of the People's Republic of China' could never happen, as the PRC had never previously had any rights or official representation in the UN.

Speaking to reporters after the vote, George H. W. Bush who, several months earlier, had been appointed by Richard Nixon as the US ambassador to the UN, made a prescient statement. 'The United Nations today crossed a very dangerous bridge,' a sombre Bush said in his Texas-via-Massachusetts drawl.

Since that fateful vote, China has steadily increased its influence over the UN and its associated organizations, such as the World Health Organization (WHO), the International Civil Aviation Organization and UNESCO. Taiwan's lack of UN representation has been a gift that keeps on giving for the CCP, which claims to speak for Taiwan in the General Assembly and its related organizations just as it does for Tibet, East Turkestan (better known by its CCP-given name, Xinjiang) and, more recently, Hong Kong. As such, Taiwan's absence from the UN is an indispensable asset – and the UN itself is increasingly helpful in isolating and erasing Taiwan by speaking as if Taiwan is indeed a part of the PRC. It increasingly defers to Beijing on matters of Taiwan's participation in both the General Assembly and its specialized agencies, to the detriment of both Taiwan and the world at large. Taiwan's international status has consistently been under attack as governments, corporations and individuals erase Taiwan's sovereignty and its people to stay in Beijing's good graces.

As authoritarian China builds up its capacity for a war of conquest against its democratic neighbour that does not threaten it, the UN's hypocrisy regarding Taiwan becomes more and more apparent. Article 1 of the UN Charter specifically mentions the 'principle of equal rights and self-determination of peoples', while Article 2 bars the 'threat or use of force in international relations'. Regardless, the viewpoint of an increasingly aggressive Beijing is what informs almost all UN discussions of Taiwan.

In March 2023, Tsai Ing-wen passed through New York before heading onwards to visit countries in Central America who still recognize the ROC government. A press briefing at the UN headquarters in Manhattan turned contentious as journalists pressed Stéphane Dujarric, spokesperson for Secretary-General António Guterres, regarding Taiwan's exclusion. Irish journalist Yvonne Murray of RTÉ News, who had previously worked as a correspondent in Beijing, asked Dujarric a straightforward question, eliciting replies that highlight the UN's complicity in propagating the PRC's fabrication that Taiwan's status was determined by the UN in 1971:[2]

MURRAY: *So, [the] Secretary-General is clearly seen as a champion of democratic values. Given that the President of Taiwan, Tsai Ing-wen, is in the US this week, will come to New York*

> later in the week and is the leader of what's considered Asia's leading democracy, does the Secretary-General have any message for her?
>
> DUJARRIC: The Secretary-General's position on China is guided by the relevant General Assembly resolution on the one-China policy.
>
> MURRAY: Sorry. I'm not asking about China, I'm asking about Taiwan and its democracy . . .
>
> DUJARRIC: No, no, I understand, and that's the answer to your question.
>
> MURRAY: OK. One other question, then. No message for President Tsai Ing-wen, but what about the Taiwanese citizens, the passport holders who are not even allowed into this building to take a tour? Does the Secretary-General have anything to say to that?
>
> DUJARRIC: The policy of the UN is that the premises of UN Headquarters are open to people with identifications of Member States of the UN.

Following up, another journalist asked whether the UN considered Taiwan to be part of China. Avoiding even saying the word 'Taiwan', Dujarric replied: 'Our position on China is guided by the General Assembly resolution passed in 1972 or 1973 on the one-China policy.' However, the resolution was passed in 1971 and never mentions Taiwan or any one-China policy.

The ROC's expulsion from the UN was not a foregone conclusion. Multiple times, Richard Nixon and Henry Kissinger had privately met with ROC officials, pushing a US proposal for dual representation – meaning the ROC and the PRC would both be recognized in the UN.[3] But the ever-proud Chiang would rather leave the global body his government had co-founded than settle for a Taiwan seat. Just before the 1971 vote, the ROC representative to the UN, Liu Chieh, spoke from the rostrum, announcing that his government would withdraw from the proceedings of the Assembly. He then led the ROC delegation from the chamber.[4]

As the historian James Lin from the University of Washington told me, 'Chiang was adamant that the ROC was the only legitimate government representing China, and he looked down upon the international organization as of secondary importance, a mere tool to further his real end of military strength for the ROC.'

Chiang Ching-kuo, who was positioned to inherit the ROC and Taiwan from his father, was less cavalier. He sought to broker a deal with Washington to secure a Taiwan seat for the ROC, but his hopes were dashed.

'The repeated votes on the ROC's status were perceived as humiliating and demoralizing by the elder Chiang, so he ordered the withdrawal of the ROC from the UN before the dual recognition plan could be voted upon, in essence allowing Resolution 2758 to pass,' Lin said. 'Chiang's obstinacy singlehandedly doomed Taiwan to its current fate of being excluded from the UN, which is a tragedy since he was an authoritarian strongman who did not represent the democratic will of Taiwanese people.'

In the more than five decades since claiming the China seat, Beijing has found increasing success in deceptively wielding Resolution 2758 as proof that the UN General Assembly endorsed it as the rightful ruler of Taiwan. But the language of the resolution clearly does not do that, and pushback is starting to build – albeit still muted from political leaders worldwide. Heads of state of non-democratic countries tend to side with Beijing on the issue of Taiwan, given their shared values, or lack thereof. And leaders of democracies are often reluctant to speak up strongly on Taiwan's behalf, as they fear economic and political reprisals from Beijing that could jeopardize their chances in future elections. Diminished access to the world's second-largest economy is hardly something one can boast about when trying to get re-elected.

Parliaments and legislatures in democracies are different from executive branches in this regard, however. They tend to be more willing to stand up to China on principles. In July 2024, the Inter-Parliamentary Alliance on China (IPAC) convened in Taipei for the first time and admitted Taiwan to its ranks. The IPAC is a coalition of lawmakers that works to reform how democratic countries approach China. The Taipei summit was the biggest parliamentary delegation to ever visit Taiwan, bringing forty-nine lawmakers from twenty-four legislatures across five continents to the diplomatically isolated capital.

During three days of meetings, delegates were addressed by both President Lai Ching-te and Vice President Hsiao Bi-khim. Taiwan was admitted as a member, with delegates present from the DPP and the TPP, a newer opposition party. The KMT was not represented. One of the major action points to come out of the summit was the 2758 Initiative, in which members vowed to return to their legislatures and push for resolutions rejecting China's distortion of Resolution 2758.[5] It quickly bore fruit.

Three weeks after the Taipei summit, the Australian Senate voted unanimously to pass an urgency motion rejecting China's twisting of Resolution 2758 and international law regarding Taiwan's sovereignty. Senator Claire Chandler said of the vote, 'This is of great concern for peace and security in our region. We must be clear-eyed about the Chinese government's stated intentions and their aggressive and dangerous behaviour.'[6]

One month later, the House of Representatives of the Netherlands – the country that first colonized Taiwan 400 years earlier – voted overwhelmingly to pass a resolution stating that Resolution 2758 does not confer PRC sovereignty over Taiwan, nor does it preclude Taiwan's participation in the UN or other international bodies. Of the House's 150 members, 146 voted in support of Taiwan.[7] Similar resolutions were subsequently passed by the parliaments of both the European Union and Canada. In

2023, the US House of Representatives had passed a similar bill, the Taiwan International Solidarity Act which, at the time of writing, was awaiting Senate approval.[8] Embarrassingly for Taiwan, attempts to pass a similar resolution in its own legislature in October 2024 were foiled by KMT opposition.[9]

Back during the Cold War, both Mao and the Chinese Foreign Minister Zhou Enlai knew that Resolution 2758 left Taiwan's status undetermined. China's relative weakness in the aftermath of the Cultural Revolution that ended in 1976 meant that resolving the 'Taiwan question' would have to be delayed. It wasn't until the 1990s that China's diplomatic and propaganda organs set themselves to work on internationalizing what it calls the One China Principle, which is:

> *There is but one China in the world, Taiwan is an inalienable part of China, and the Government of the People's Republic of China is the sole legal Government representing the whole of China.*

With the end of the Chiang Dynasty in the 1980s and the subsequent democratization of Taiwan in the 1990s, the ROC claim to be the legitimate government of both sides of the strait faded. In the early 1990s, Lee Teng-hui abandoned the Chiang's designs to 'retake the mainland', a move that deprived Beijing of a key piece of propaganda against Taiwan. This effectively meant that from then on, only one side of the Taiwan Strait – China – was preparing to invade the other. When international media speaks of 'rising cross-strait tensions', it often fails to make explicit that whether tensions rise or fall is decided by Beijing alone.

For China, losing the antagonism of the ROC increased the importance of muddying the waters around the UN's 1971 resolution and blurring the lines between individual nations' one-China policies and the One China Principle. One-China policies are decided by individual states, and can, like the US, note the existence of China's claim on Taiwan without officially approving of it, or, like Russia, fully support and endorse Beijing's One China Principle, as described above. China has signed secret agreements with UN bodies regarding Taiwan, successfully restricting Taiwanese access to the UN and its facilities, and has worked to great effect to place Chinese nationals in staff positions high and low across various departments.[10] It has also succeeded in pushing the UN to adopt CCP-approved wording when referring to Taiwan, such as 'Taiwan, China' or 'Taiwan, Province of China'.

The WHO is a prime example of this influence. In 2008, following the election of Ma Ying-jeou, and his endorsement of the 1992 Consensus, Beijing allowed Taiwan to have observer status at the WHO's annual World Health Assembly (WHA). Although the WHO Secretary-General has the power to invite Taiwan to the WHA as an observer, the agency has consistently deferred to Chinese sensitivities. I obtained multiple faxes from the PRC permanent mission in Geneva to the Office of the UN High Commissioner for Human Rights, two of which highlight that Taiwan's participation is a matter that Beijing decides, and that decision is based not on concern for global health, but on the rejection of Taiwanese self-determination – a major area of agreement for the CCP and KMT.

One fax sent on 19 May 2016, the final day of Ma's presidency, paints a picture of Beijing as charitable and magnanimous. The fax stated:

> *The participation by the Taiwan Province of China as an observer in the annual World Health Assembly since 2009 has been a special arrangement based on the One-China principle. As a sign of goodwill, this special arrangement demonstrates the sincere wish of the Chinese Government to maintain the smooth and peaceful development of the ties across the Taiwan Straits, and to improve the well-being of our Taiwan compatriots in the field of public health.*

This wording makes clear that China is using a global health body as leverage over Taiwan's domestic politics.

The day after that fax was sent, Tsai Ing-wen was inaugurated as president. In her speech to the nation, she expressed her hope for talks with China but emphasized that having just sworn to uphold the ROC constitution, she could not enter into any discussions in which renouncing sovereignty was a precondition. Several days later, Taiwan participated in the WHA. It has not been allowed back since.

In May 2017, with that year's WHA meeting just around the corner, the PRC mission sent another fax, which read:

> *The Permanent Mission of the People's Republic of China to the United Nations Office at Geneva and Other International Organizations in Switzerland (hereinafter referred to as Permanent Mission of China) presents its compliments to all International Organizations in Geneva and has the honor to inform the latter that the Chinese Government has decided that Taiwan Province of China shall not participate in the 70th World Health Assembly to be held from 22 to 31 May 2017 in Geneva, Switzerland.*

Every year since, the US, sometimes joined by other countries, has advocated Taiwan's participation as an observer in the WHA. It did so again in May 2024, but was unsuccessful – an almost foregone conclusion, given that Taiwanese voters had not voted the KMT back into power earlier in the year.

Bonnie Glaser, managing director of the Indo-Pacific Program at the German Marshall Fund, told me that even though Washington doesn't officially recognize Taiwan's government, that doesn't preclude it from pushing for Taiwanese inclusion in the WHA, even if only as an observer rather than a full member.

'If the US were supporting Taiwan's membership in the UN as a sovereign state, then there would be a contradiction,' Glaser said. 'But US policy is clear – it supports meaningful participation for

Taiwan in international organizations that require sovereignty for membership.

'There is no legal basis for barring Taiwan's participation in the UN and its specialized agencies,' she emphasized. 'United Nations General Assembly Resolution 2758 was silent on that question – I would argue that Beijing also holds that view since it allowed Taiwan to participate as an observer in the WHA [during the Ma era].'

Momentum in favour of Taiwan's involvement in the WHA has been growing outside of Washington. In May 2023, the unofficial diplomatic missions of the US, Japan, Germany, UK, Australia, Canada, Czech Republic and Lithuania in Taipei released a joint call for Taiwan's 'meaningful engagement' with the WHO and participation as an observer in the WHA.[11] Saying Taiwan faces an uphill battle to participate in the UN or its specialized agencies in any capacity would be an understatement. But occasionally cracks emerge in the UN's stance on Taiwan.

In September 2023, Deputy Secretary-General Amina Mohammed, the organization's second highest-ranking official, expressed support for resolving the issue of Taiwan's status within the body.[12] During a briefing about the UN Sustainable Development Goals, Mohammed was asked by *National Review* reporter Jimmy Quinn for her thoughts on the exclusion of Taiwan – a near G20 economy and significant polluter.

'I think [the] exclusion of anyone holds back the goals. We said leave no one behind, and I think that the states have to find a way to make sure that we are not in that position where we're excluding people,' she said. 'Every person matters, whether it's Taiwan or otherwise, and I think it's really important for member states to find the solution to that,' Mohammed added.[13]

To the surprise of no one, when deputy UN spokesperson Farhan Haq was asked about Mohammed's statement just an hour later at a separate event, he answered as if Taiwan was part of China: 'We don't intend to leave any of the people of China behind, and we support all of the people of China, but we stick by the one-China policy as has been decided by the General Assembly,' Haq said, declining, as many of his colleagues also choose, to even utter the word 'Taiwan'.

It has become clear that Beijing is seeking to conflate the one-China *policies* of individual and autonomous actors with its own One China *Principle*. The US and most other democracies broadly acknowledge China's claim, without recognizing it. Even if they recognize the PRC and not the ROC government in China, they still have their independent unofficial relations with Taiwan, with varying degrees of closeness.

In the late 1990s, as many as thirty countries recognized the ROC government in Taiwan – but that was during the era in which these so-called 'diplomatic allies' were far more important in domestic political discourse than they are today. Most Taiwanese people cannot find the countries that have official ties with their government on a map, and most have little interest in their government using chequebook diplomacy to keep the remaining countries onboard. Smaller countries over which Beijing holds major leverage, however, are increasingly used to promote the line that Taiwan is Chinese territory. They have recently begun to also advance the false narrative that the UN decided Taiwan's status in 1971.

In January 2024, just two days after Taiwanese voters elected Lai Ching-te their next president, the Pacific Island nation of Nauru broke ties with Taipei to establish official diplomatic relations with Beijing, leaving only eleven countries and the Holy See recognizing the ROC. Nauru's government said that it was recognizing China as Taiwan's government in order to be compliant with UN Resolution 2758. Following Nauru's announcement Laura Rosenberger, chair of the American Institute in Taiwan, described Nauru's decision as 'unfortunate'.

'While the government of Nauru's action is a sovereign decision, it is nonetheless a disappointing one,' Rosenberger said. 'UN

Resolution 2758 did not make a determination on the status of Taiwan, does not preclude countries from having diplomatic relationships with Taiwan and does not preclude Taiwan's meaningful participation in the UN system,' Rosenberger said. 'It is disappointing to see distorted narratives about UN Resolution 2758 being used as a tool to pressure Taiwan, limit its voice on the international stage and limit its diplomatic relationships.'[14]

Domestically, diplomatic recognition means different things to Taiwan's main parties. For example, for the DPP, unofficial allies and friends (such as the US, Japan, the UK and, more recently, Central and Eastern European states) are incredibly important to Taiwan's current political and economic situation. In practical terms, these allies are much more likely to be of use if China attacks than ties with Paraguay, Haiti or Saint Lucia. For the KMT, who views Taiwan through the lens of the ROC party-state it brought over in 1945, each loss of official recognition is viewed as a blow to the ROC's legitimacy.

If China keeps poaching the remaining countries, it also brings up an interesting question once posed to me by a smiling Taiwanese diplomat: 'If nobody recognizes the ROC any more, does it just become Taiwan?' But how the UN treats Taiwan does matter, not just in a participatory sense, but also in terms of the knock-on effects outside the UN ecosystem.

In nearly a decade of writing about Taiwan, I've been told by multiple editors at different publications that the reason I'm not allowed to refer to Taiwan as a country is because it's not recognized by the UN.

Websites for companies and organizations around the world often take their lists of countries from the International Organization for Standardization (ISO). Headquartered in Geneva, the ISO oversees more than 25,000 standards that are part of the fabric of global trade and commerce. Among them are 'country designations', which are used globally as accepted standards for how to refer to the world's sovereign states. In early 2024, while conducting online research for this book in Taipei, a small window entitled 'Set your location' popped up on a website I had just accessed.

'It looks like you're in Taiwan, Province of China,' the pop-up window told me. 'Would you like to update your location?'

A dropdown menu was encouraging me to select 'Taiwan, Province of China', without offering a 'Taiwan' option. It was not the first time I had seen Taiwan presented as part of China on a non-Chinese website. Since 2018, Beijing has substantially stepped up pressure on all international companies and organizations with a presence in China to harmonize their websites' descriptions of Taiwan with its One China Principle.[15]

I contacted the company that owned the website, who told me a third-party vendor managed the website for them, and had been using ISO 3166-1 to supply user location names. With that information, I then contacted ISO itself, which told me that the source of the country names in ISO 3166-1 was none other than . . . the UN.

For my friend Jenny, who, like many children of Taiwanese immigrants in the US, was raised to spread the word regarding Taiwan's autonomy and unique identity, the erasure of Taiwan's sovereignty is personal. Her parents emigrated from Taiwan to the US in 1990, and most of her extended family still resides in Taiwan. Whenever her relatives would visit her in New York, they would be unable to go on tours of the UN like the one we had taken in 2019. This is, of course, because the UN does not allow anyone into any of its buildings on an ROC passport. The irony is that globally, 141 countries allow visa-free entry to Taiwan's ROC passport holders, compared with 81 countries who do so for China's PRC passport holders.[16]

'Having gone to Taiwan many times while growing up, I knew, saw, and felt how Taiwan is just like any other country – nothing felt abnormal about it,' she told me. 'Yet, as I grew up and learned more about Taiwan, I started to see how the rest of the world views Taiwan – as abnormal and, sadly, as non-existent.'

This is more than a problem for Taiwanese, however. In early 2020, Taiwan's gold-standard response to the initial outbreak of Covid-19 was the envy of much of the world. As lives were lost and economies devastated by the pandemic, Taiwan managed to survive for two years without any major outbreaks – or any lockdowns. Had Taiwan been a WHO member, its doctors' experience with coronaviruses from SARS in 2003 and their familiarity with the Chinese government's obfuscation and obstructionism could have improved the global response during the early days of the pandemic, preventing untold deaths and economic damage in the process.

Because the UN is blocking Taiwan's participation in numerous organizations at the behest of the CCP, the world is denied the benefits of a modern, highly open and accountable democracy that has much to contribute. This is unlikely to change with the next global health crisis.

From Allies to Unofficial Friends

It was the end of December 1978, less than one week before the US would officially sever diplomatic relations with the ROC to recognize the PRC. Deputy Secretary of State Warren Christopher's plane touched down on the tarmac at Taipei's Songshan Airport, where he was greeted by Ambassador Leonard Unger and ROC Foreign Minister Frederick Chien. Upon his arrival, Christopher announced to local media that he sought to establish ways 'to maintain our cultural, commercial and other relationships on an unofficial basis'. Striking an ominous tone, Chien told Christopher, 'During your stay in Taipei, you will gain a clear understanding of the position of our government, and the feelings of our people.'[17]

The feelings of the Taiwanese people became clear as the

US delegation left the airport. Hundreds of angry, yelling Taiwanese – some holding signs saying 'Carter is a fool' or 'Carter, go home to sell your peanuts' – lined the narrow street leading out of the airport. Hurling verbal abuse at the vehicles, they pelted the windows with eggs and tomatoes and eventually forced the cars to stop, with police unable, or perhaps unwilling, to help. Balls of mud containing rocks followed. Some protestors used bamboo poles to break car windows before police restored order and the cars made their way into the city. There were no serious injuries; Christopher's face was cut and his glasses were broken.[18]

Journalist Andrew Nagorski, who had just arrived from Hong Kong to cover the American delegation's arrival for *Newsweek*, had the fortune – or misfortune – of catching a ride with the CBS crew. They had rented a black car that looked very much like the cars that Taiwan's foreign ministry had provided the American delegation.

'The CBS crew told the driver to get as close to the Christopher motorcade as he could, and he was so eager that he whipped in front of all the cars and our car ended up at the front, which meant we looked like the lead car,' he said.

As soon as the line of cars began making its way away from Songshan Airport, the crowds assembled outside surrounded the CBS car, he said, pelting it with eggs, pounding on the windows and windshield, and bringing them to a halt. Unable to clearly see what was going on, Nagorski and the others became increasingly concerned for their safety.

Being in the press bus at the rear of the line wasn't much better. Another American journalist, Don Shapiro, was a passenger on the bus. As he recounts, shortly after the angry crowd broke one of the windows near him, he was hit by a flying ham sandwich.[19]

During this incident, law enforcement officers seemed to have vanished. While most of the on-the-ground reports suggested that the protest was government-organized and the light police touch was intentional, there was also speculation that turnout was larger than expected, catching the police and government off guard.

'The police, usually so prevalent in that era, had clearly made themselves scarce,' Nagorski told me. 'It took what felt like a long while before they must have cleared a path for us, allowing us to leave.'

Such a massive protest was extremely rare during martial law – it seems unlikely to be a coincidence that it happened to serve the KMT government's political goals. The US State Department registered a 'very strong protest',[20] after which the ROC government expressed regret and offered assurances that the US party would be safe during the remainder of its visit. Such had become the state of ties between the decades-long allies.

For years, speculation had been steadily growing that, one day, the ROC's derecognition by the US would come. Despite this, there had been no prior discussion between the US and ROC regarding the extent of their unofficial relations. The reasons for this lack of preparation were understandable. Carter's White House and State Department didn't want to cause alarm among ROC diplomats, who might rally their many friends in Congress to take pre-emptive action that could complicate the recognition of Beijing. Inside the foreign ministry in Taipei, those who wanted to broach the subject with the Americans in order to get the best deal possible were shushed by their superiors. Chien would later recount his boss, Foreign Minister Shen Chang-huan, likening such discussions to 'putting the lid on your own coffin'.[21] To facilitate continued US–Taiwan interactions following the ROC's derecognition, a daunting slate of issues needed immediate attention. This included travel, trade and ROC assets in the US. The Mutual Defense Treaty would cease, raising the questions of whether Taiwan would be covered by the American security umbrella and whether the ROC military could purchase defensive articles. At the core was the basic issue of how the governments would interact going forward.

When Chien met with Christopher the day after the motorcade incident, he asked to see the draft legislation he presumed the US had prepared. The scraped-up American diplomat elicited further displeasure from Chien when he informed him that not even a draft proposal existed. Days later, Christopher would return to Washington, having accomplished little more than being a scapegoat for a new nadir in US–ROC relations.

As with the UN's expulsion of the ROC, the ending of official government relations between the US and the ROC governments was a result of Chiang obstinacy in the face of unpleasant realities. After Chiang Kai-shek died in 1975, Chiang Ching-kuo continued to cling to the notion that the ROC government was the sole legal government of all of China, which included Taiwan as a province. It didn't matter to the Chiangs that the ROC had become an isolated rump state that only administered Taiwan and the Fujianese island counties of Kinmen and Matsu. It was unquestioned in Taipei that the ROC would one day rightfully retake the motherland.

Had the Chiangs been willing to accept a two-state solution in which they relinquished their claim to represent China in exchange for a seat at the UN under the name of Taiwan, a path towards US–Taiwan recognition might have been possible. The PRC would have protested, but it had much less leverage then than it does today. The Chiangs' unwillingness to abandon their China dream helped trap Taiwan in the diplomatic limbo which it inhabits to this day.

The History of US Policy Towards Taiwan

The US's current-day Taiwan policy is a patchwork of very careful language spread across one federal law, three communiqués with China and a once-classified set of six assurances to Taiwan made by President Ronald Reagan. These main tenets of the US approach to

Taiwan are couched in unnatural wording as well as references to American ties with China. At the core of this policy are two main pillars:

1. Taiwan's status remains undetermined.
2. It is up to the people of Taiwan to determine their future, free of coercion.

At the beginning of 2024, US Secretary of State Antony Blinken issued a statement of congratulations to Lai Ching-te on his victory in Taiwan's presidential election:

The United States congratulates Dr. Lai Ching-te on his victory in Taiwan's presidential election. We also congratulate the Taiwan people for once again demonstrating the strength of their robust democratic system and electoral process.

The United States is committed to maintaining cross-Strait peace and stability, and the peaceful resolution of differences, free from coercion and pressure. The partnership between the American people and the people on Taiwan, rooted in democratic values, continues to broaden and deepen across economic, cultural, and people-to-people ties.

We look forward to working with Dr. Lai and Taiwan's leaders of all parties to advance our shared interests and values, and to further our longstanding unofficial relationship, consistent with the US one China policy as guided by the Taiwan Relations Act, the three Joint Communiqués, and the Six Assurances. We are confident that Taiwan will continue to serve as an example for all who strive for freedom, democracy, and prosperity.[22]

At first glance, this statement might seem to be a run-of-the-mill congratulatory message. Small details, however, indicate a deferential tone towards Beijing. For example, Blinken refers to Lai as 'doctor' rather than 'president-elect'. The rather awkward usage of 'the Taiwan people' and 'the people on Taiwan' avoids recognizing the existence of the very real Taiwanese nationality. After these concessions, Blinken invokes the Taiwan Relations Act, three Joint Communiqués and Six Assurances. While most people who aren't focused on US relations with China and Taiwan might find themselves suddenly lost in diplo-jargon, these documents could be summarized roughly like this:

> *The US only recognizes the government in Beijing as the legitimate government of China and duly notes that government's claim on Taiwan, without recognizing it. The status of Taiwan remains unresolved and the US will accept however that status is resolved, provided it is peaceful and with the consent of the Taiwanese people. To that end, the US will provide Taiwan with whatever it needs to defend itself from non-peaceful coercion.*

The piecemeal assembly of unofficial American post-war policy towards Taiwan began with the Shanghai Communiqué of 1972, also known as the First Joint Communiqué, which was issued at the end of the landmark meeting between Nixon and Mao. It was signed at the Jinjiang Hotel in Shanghai's old French Concession on 28 February – long a cursed date in Taiwan. The unofficial summit and communiqué came less than a year after the ROC's ejection from the UN, adding momentum to the growing international isolation of the Chiang regime, and by extension, the Taiwanese people.

After running through a laundry list of disagreements on geopolitical issues elsewhere in Asia, the ninth point of the communiqué notes that 'progress toward the normalization of relations between China and the United States is in the interests of all countries', with both sides stating shared goals of peace and rejecting regional hegemony.

Point 11 discusses Taiwan, with China first offering its take:

> *The Taiwan question is the crucial question obstructing the normalization of relations between China and the United States; the Government of the People's Republic of China is the sole legal government of China; Taiwan is a province of China which has long been returned to the motherland; the liberation of Taiwan is China's internal affair in which no other country has the right to interfere; and all US forces and military installations must be withdrawn from Taiwan. The Chinese Government firmly opposes any activities which aim at the creation of "one China, one Taiwan", "one China, two governments", "two Chinas", an "independent Taiwan" or advocate that "the status of Taiwan remains to be determined".*[23]

In the following point, the US declared:

> *The United States acknowledges that all Chinese on either side of the Taiwan Strait maintain there is but one China and that Taiwan is a part of China. The United States Government does not challenge that position. It reaffirms its interest in a peaceful settlement of the Taiwan question by the Chinese themselves. With this prospect in mind, it affirms the ultimate objective of the withdrawal of all US forces and military installations from Taiwan. In the meantime, it will progressively reduce its forces and military installations on Taiwan as the tension in the area diminishes.*

The core of this communiqué is the US *acknowledging* that the PRC and ROC governments maintain that Taiwan is part of 'one China' while not formally *recognizing* this claim.

For the US, Taiwan's status was and remains undetermined. However, when Henry Kissinger and the Chinese Foreign Minister Zhou Enlai met secretly in Beijing in July 1971, Kissinger assured Zhou that the US neither supported 'two Chinas' or 'one China, one Taiwan', while also vowing to not support the Taiwanese who sought to overthrow Chiang.[24] The State Department had also temporarily ceased to mention Taiwan's undetermined status. The substance of the talks between Kissinger and Zhou overlooks the fact that neither party had any authority to speak on behalf of the Taiwanese people – this was pure big-power horse-trading.

During his visit to China with Kissinger in 1972, Nixon made several comments in his discussions with Zhou that suggested that he hoped that, after normalization of US–PRC ties, the unification of Taiwan and China would come about naturally over time. Zhou expressed concern multiple times over Japanese forces returning to Taiwan, to which Kissinger replied: 'We will oppose this to the extent that we can control Japan.'[25]

Chiang Kai-shek may have taken some comfort in the Watergate scandal putting China diplomacy on the back burner for the Nixon administration but, after the communiqué, it was apparent that the eventual abandonment of the ROC by the Americans was only a matter of time. That time came in December 1978, three years after Chiang's death, with Carter's announcement of mutual recognition between the world's most powerful country and the world's most populous country. This would happen on 1 January 1979 with the Joint Communiqué on the Establishment of Diplomatic Relations, generally known as the Second Joint Communiqué.

A ceremony with full military honours and a nineteen-gun salute welcomed Deng Xiaoping to the White House on 29 January. That night, he was the guest of honour at a state dinner attended by Nixon – Nixon's first time in his old residence since resigning in disgrace in 1974. Carter and Deng ditched Nixon afterwards for a night at the Kennedy Center, where they took in performances by the Joffrey Ballet, John Denver and Shirley MacLaine. While he was in Washington, Deng signed the second communiqué. He would next head down to Georgia and sign deals bringing Coca-Cola and Ford Motor Company into the Chinese market. He then visited a rodeo in Texas, where his donning of a Stetson cowboy hat immediately endeared him to a generation of Cold War-weary Americans. A final stop in Washington state to buy some Boeing 747s and catch up with 'old friend of the Chinese people'[26] Henry Kissinger rounded out the first official visit by a PRC leader to the US.

For Chiang Ching-kuo, the reception that the Carter administration rolled out for Deng and the PRC must have been humiliating. Decades before, Chiang and Deng were classmates at Sun Yat-Sen University in Moscow. Later, troops loyal to Chiang's father would chase Deng and 100,000 Communists out of southern China's Jiangxi province, from which they set off on the gruelling Long March. After a year's progress, the march ended in China's north with less than 10,000 survivors. One of those survivors was Deng.

In terms of substance, the second communiqué reaffirmed the principles of the first, with the addition of the US recognizing the PRC as the sole legal government of China. The US continued to acknowledge the fact that Beijing claimed Taiwan as PRC territory, while still not officially recognizing that claim . . . at least in the English-language version of the joint statement. It was in the Chinese translation of the communiqué that the PRC eked out a small win, using the verb *cheng ren* (承认) to describe the US position on the Chinese claim on Taiwan, which in Mandarin means 'recognize'. One of the participants behind the scenes said in an interview two decades afterwards that he was certain that political expediency led the State Department to overlook the obvious change made to the language in the Chinese version.

In 1977, Harvey Feldman was made the director of the Office of Republic of China Affairs with the challenging remit of managing the future unofficial relationship with the ROC. Feldman was fluent in Mandarin and had previous experience in Taiwan and Hong Kong. In a 1999 interview, Feldman accused fellow diplomat J. Stapleton Roy of knowingly approving the contradictory language in order to avoid causing any delay to normalization. In Feldman's words:

> *The PRC tried to pull a fast one; in their Chinese version of our communiqué, they used a two-character phrase cheng ren (or 'recognized'). This a phrase that is used when speaking of a recognition of a government. When the two-character phrase appeared in the PRC text, the US Liaison Office in Beijing should have immediately expressed its disapproval of the PRC text. [. . .] J. Stapleton Roy, at the embassy at the time, was someone who had been born in China, had grown up in China and was therefore bilingual; he had served in Chinese-language positions for most of his career. He was completely aware of the difference between the two phrases.[27]*

Feldman continued:

> *Roy should have immediately pointed out that the PRC had mistranslated the American text, but I am firmly convinced that because he was so keen to achieve 'normalization,' he did not point out this very important mistranslation. So the official Chinese text of the communiqué included the phrase cheng ren ('recognized'). This has created no end of mischief in the PRC–US relationship because in effect the two versions of the same communiqué say different things. The Chinese version uses the word 'recognition' of the PRC claim of one China, whereas the English version says 'the US acknowledges the PRC contention.'*

Regardless of the circumstances behind Roy's handling of the term *cheng ren*, the deal had been done: the ROC (and Taiwan) had been jettisoned for the PRC.

By the time a triumphant Deng was flying back to Beijing, however, Congress had sprung into action. Carter's announcement had caught Capitol Hill off guard, with many lawmakers furious at being left out of the loop. Congress urgently wanted to solve the issue of how to move forward with Taiwan as an unofficial ally. The solution to this tricky political problem largely came from the private sector. Given the significant size and scope of American investment in and trade with Taiwan, the Taipei office of the American Chamber of Commerce (AmCham) had already assembled a post-derecognition policy paper. AmCham Taiwan chairman Robert Parker became the face of the campaign to ensure a smooth transition for future US–Taiwan relations. In February 1979, Parker travelled to Washington, where he spoke with both houses of Congress. His hearing would air on PBS in the US and be rebroadcast by Taiwan's three networks.

Just months before Carter's bombshell announcement, the Senate had passed a resolution stating that no action should be taken regarding the Mutual Defense Treaty with the ROC without first consulting Congress. By shunting that aside, Carter had insulted the legislative branch which, Parker noted in a later interview, became very receptive to Parker and AmCham's suggestions.

Testifying in Congress in a grey suit and burgundy tie, Parker made the case for maintaining trade and legal ties with Taiwan. He wanted to ensure minimum disruption to US–Taiwan trade at whatever cost. This included the US guaranteeing Taiwan's security, as it had before derecognition. 'Part of doing business, and doing it successfully, is, having certainty,' Parker said. 'Business thrives on certainty. And no element of certainty is more important than one's physical security. And so obtaining a strong security resolution from this committee, from the Congress, is important.'

Parker noted that he thought the defensive provisions for Taiwan could be improved upon. Up to that point, the US had only discussed getting involved in Taiwan's defence in the case of a direct attack on ROC-controlled territory by the PRC. 'I think a naval blockade of Taiwan is a real threat, in terms of what military action might be taken against Taiwan by the PRC, more so, I would submit, than an amphibious assault on Taiwan itself,' Parker said, matter-of-factly.

The language of the post-derecognition road map that had been assembled by Congress at that juncture would not address the threat of a blockade. Parker also advocated the sale of cutting-edge defensive weaponry to Taiwan so as to mitigate China's size advantage.

As Parker spoke, the feed from the Senate cut to the man sitting across from him: a thirty-six-year-old senator from Delaware named Joseph R. Biden. Biden was watching intently and taking notes. More than four decades later, as US president, Biden would answer reporter questions about whether American troops would help defend Taiwan if it were attacked by China. Each time, he replied, without hesitation: yes, they would. Biden was no stranger to Taiwan; he was one of the signers of the Taiwan Relations Act of 1979, which today is the most important and detailed of the main documents that guide American policy towards its unofficial friend.

As Parker made his case for ensuring that all agreements entered into with Taiwanese partners were backed by the force of law, he argued that, aside from the derecognition of the government in Taipei, the US government and legal system should continue to treat Taiwan as a sovereign entity.

Parker carefully separated the people of Taiwan and the ROC government on Taiwan in his remarks. He concluded by calling for legislation that would ensure a continued relationship with Taiwan that would benefit US business interests, as well as the people and government in Taiwan:

> *A prompt enactment of this legislation is not only in the interest of American business, and its trade and investment with Taiwan, but is also necessary, I think, to satisfy the moral commitment of our country to the people of Taiwan – and to a government which has been our loyal ally.*

In March 1979, the House passed the Taiwan Relations Act (TRA) by a vote of 345 to 55, followed by a 90 to 6 vote in the Senate. These two majorities would override any veto by Carter, who had little choice but to sign the bill into law on 10 April. In a clear middle finger raised by Congress to the White House, the TRA 'restored almost all the attributes of statehood to Taiwan in its relationship with Washington, and committed the United States to a permanent interest in Taiwan's security through the sale of defensive weapons'.[28] In addition to addressing legal, business and defence concerns, the TRA also established the American Institute in Taiwan (AIT). Today, the AIT is a prestigious assignment for US diplomats, unofficially performing all of the functions of an embassy.

The TRA may have been a bittersweet win for Taiwan's government and people, but its passage didn't prevent relations between Washington and Beijing from warming quickly in the following years. The same month that the TRA was passed, Biden led a Senate delegation to Beijing where he met with Deng Xiaoping. Biden convinced the Chinese leader to let the CIA build a listening post in Xinjiang to monitor Soviet arms treaty compliance. The two countries shared the intelligence gathered at the station, leveraging American technology and Chinese geography.[29]

Carter achieved his goal of normalizing ties between the US and the PRC, but that wasn't enough to keep him from losing to Ronald Reagan in the 1980 presidential election. The former movie actor and governor of California had vowed to be tough on the PRC, making him the first candidate to engage in what would become a quadrennial American tradition.

Once in office, Reagan softened up considerably. He sought to use China as a counterweight against the Soviets, and Beijing was more than happy to comply. During the Reagan era, Beijing provided weapons to the Mujahideen in Afghanistan on Washington's behalf and also armed the Contras in Nicaragua. The warming of ties between the two countries continued.[30] Only half a year after taking office, Reagan's Secretary of State Alexander Haig announced in Beijing that the administration had decided in principle to begin selling arms to China.

A year later, on 17 August 1982, Washington and Beijing issued their Third Joint Communiqué. This communiqué touched upon arms sales to Taiwan, an issue from previous meetings that had been kicked down the road. China reiterated its claim on Taiwan and its opposition to US arms sales to the ROC. The communiqué also appeared to put Washington on the road to eventually ending arms sales to Taiwan – a goal of Haig's, who, like Kissinger, was all-in on ties with China.

Among key lines that raised eyebrows in Taipei was Point 6, which stated:

The United States Government states that it does not seek to carry out a long-term policy of arms sales to Taiwan, that its arms sales to Taiwan will not exceed, either in qualitative or in quantitative terms, the level of those supplied in recent years since the establishment of diplomatic relations between the United States and China, and that it intends gradually to reduce its sale of arms to Taiwan, leading, over a period of time, to a final resolution. In so stating, the United States acknowledges China's consistent position regarding the thorough settlement of this issue.[31]

Point 7 may have been a bit unsettling as well:

In order to bring about, over a period of time, a final settlement of the question of United States arms sales to Taiwan, which is an issue rooted in history, the two Governments will make every effort to adopt measures and create conditions conducive to the thorough settlement of this issue.[32]

In an internal memo dated the day of the communiqué, Reagan offered his interpretation:

As you know, I have agreed to the issuance of a joint communiqué with the People's Republic of China in which we express United States policy toward the matter of continuing arms sales to Taiwan. The talks leading up to the signing of the communiqué were premised on the clear understanding that any reduction of such arms sales depends upon peace in the Taiwan Straits and the continuity of China's declared "fundamental policy" of seeking a peaceful resolution of the Taiwan issue.

In short, the US willingness to reduce its arms sales to Taiwan is conditioned absolutely upon the continued commitment of China to the peaceful solution of the Taiwan–PRC differences. It should be clearly understood that the linkage between these two matters is a permanent imperative of US foreign policy.

In addition, it is essential that the quality and quantity of the arms provided [to] Taiwan be conditioned entirely on the threat posed by the PRC. Both in quantitative and qualitative terms, Taiwan's defense capability relative to that of the PRC will be maintained.[33]

Reagan clearly had immediate buyer's remorse over the communiqué that Haig had sold him, as highlighted by a cable entitled 'Assurances for Taiwan' drafted on 17 August 1982.[34] US Secretary of State George Shultz told the director of the AIT, James Lilley, to urge Taiwan's government to make a public statement noting that:

[. . .] based on information received through appropriate channels, it is their understanding that the US side:

1. *Has not agreed to set a date for ending arms sales to Taiwan*
2. *Has not agreed to consult with the PRC on arms sales to Taiwan*
3. *Will not play any mediation role between Taipei and Beijing*
4. *Has not agreed to revise the Taiwan Relations Act*
5. *Has not altered its position regarding sovereignty over Taiwan*
6. *Will not exert pressure on Taiwan to enter into negotiations with the PRC.*

The 'Six Assurances', as they became known, were not declassified until August 2020.

*

In terms of messaging, China's approach to Taiwan is much easier to grasp. Beijing's message, however inaccurate, is simple: Taiwan is, and always has been, ours.

Contrast this with the hodgepodge American approach to its relationship with Taiwan under the umbrella of Washington's one-China policy. It's hard to sound like you have a coherent policy if you have to say that policy is 'guided by the Taiwan Relations Act, the three Joint Communiqués and the Six Assurances'. This highlights just how ad hoc and haphazard American policy towards both China and Taiwan has been since Nixon and Mao's 1972 meeting.

Shifting relations in the three-way relationship between Washington, Beijing and Taipei over the decades have also had ramifications for journalists – the people who are supposed to make sense of it all for those who are focused on other things. Although both Taiwan and the US have limited journalistic access to China today, that wasn't the case not long ago.

The Chinese Media

Not everyone is meant to be an accountant.

In the late 2000s, as a student at National Taiwan University – which could be described as the Harvard or Oxford of Taiwan – Jane Tang was beginning to accept that majoring in accounting, while a pragmatic choice, was not for her. After taking a few sociology classes on the side, she realized that she wanted to have an impact on society at large. A native of Taiwan's second-most populous city, Taichung, Tang heard about an internship at the highly reputed Chinese business magazine *Caijing*. Beijing had just wowed the globe by hosting one of the most impressive Olympics in recent memory, and its economy was booming. While her mother's side of the family had been in Taiwan for many generations, Tang's paternal grandfather was from the northern Chinese province of

Shandong and she had an interest in what things were like on the other side of the strait. With her friend's introduction, Tang landed the internship at *Caijing*, moving to Beijing in 2009 when Taiwan's new president, Ma Ying-jeou, was promoting greater cross-strait exchanges.

'That was a totally different time,' Tang told me. 'After the Olympics, there was all this hope for China – that it would open up more, there would be more understanding, more opportunities.'

For Tang and other young Taiwanese at the time, China was a welcoming, open country with a booming economy and a place where bosses and colleagues were more interested in their views and ideas than back home. I lived in China at that time and met many of Tang's countrymates who, while being deeply proud of being Taiwanese, were riding the wave of Chinese development – much of it driven by Taiwanese investment. They were doing business, making friends, and falling in love – all while exploring and experiencing China's vastness.

For the young Taiwanese cohort of that day, it was exhilarating to work in an ascendant China. This China had not yet indicated it would seize and militarize numerous island groups in the South China Sea, destroy the one country, two systems arrangement it had agreed to uphold in Hong Kong under a legally binding agreement with the UK, or militarily harass Taiwan daily with warplanes, warships and even drones and balloons.

'It was very accessible for us because it was close to Taiwan, and we speak the same language,' Tang said of China. 'My dad's father was from Shandong province – I didn't know the country, but I wanted to see what was going on.'

Like many other young Taiwanese, Tang was surprised by how much her superiors asked her for her input. Among her colleagues at *Caijing* were also a significant number of *haigui*. This is the Mandarin word for 'sea turtles', a nickname for Chinese students who had studied abroad and returned to forge a new future for a China that, at the time, felt far from the Mao era. 'They were all very open-minded,' she said of her *haigui* colleagues. 'They would ask me about Taiwan.'

Prior to crossing the strait, Tang had interned at a television station in Taiwan, where she had a very different experience. Hierarchy and seniority reigned and work was a grind rather than a vocation. 'I was just copying content from newspapers, I wasn't really learning anything,' she said. 'In China, I felt everyone in the newsroom was treated equally. People could share their views, and were eager to learn.'

During her internship at *Caijing*'s open-plan office, 'they were really trying to run a Western newsroom,' Tang said. At the beginning of her internship, she remembered hearing an editor yelling into a phone, telling the person at the other end of the line off. When she asked a Chinese colleague what the commotion was about – she was not accustomed to Beijing-accented Mandarin – the colleague informed her the editor was rejecting a government official's attempts to interfere in an article.

'One editor told us all that the government were hooligans, and that we'd have to out-hooligan them,' she told me. 'I was inspired from seeing that, the conversations and the pushback against the government. Moments of "Should we do this? Yes!" and then chasing a story. I remember thinking, "Oh my god, this is real journalism."'

As a Taiwanese who had learned the stories of how her country moved from dictatorship to democracy, the heady environment in which Tang found herself reminded her of *Formosa Magazine* and the Tangwai movement behind it – a movement that, years later, succeeded in ending the KMT's monopoly on political power in Taiwan.

'People in the newsroom wanted to do something like that,' she said. 'You're not working there for the money or anything but you don't care because you know it's a very special thing you're doing.'

Rather than the rhetoric about unification with the motherland and rejuvenation of the great Chinese nation advocated by Xi Jinping today, Tang said many people she spoke with in Beijing saw Taiwan's democratic system as more appealing in the long term. 'People would sometimes be like, "When can you guys come back to save us?",' she said, '"Because we want to be like you one day."

'I never heard anything like, "We're going to take you back one day." Maybe here or there a taxi driver might make some anti-DPP comments, but people tended to be more curious: "What's Taiwan like?" People were very willing to engage with you on any level.'

After a management shakeup in 2009, Tang and many of her colleagues joined Caixin Media. As a staff reporter, Tang covered stories about the forced demolition of neighbourhoods by the government, China's gender-equality issues and other topics that would be considered much more 'sensitive' by today's media's CCP minders. She also reported from Taipei, New York and eventually Washington, DC.

After five years, Tang left Caixin but stayed in Washington, where she was a contributing reporter for the BBC's Mandarin service for two years, before eventually returning to a Chinese employer, the tech giant Sina, in early 2015. Tang's experience covering US politics appealed to Sina, who, along with other tech companies like Baidu, sought to move into news media. Tang described that time as 'the last hopeful era for Chinese media'.

Speaking to the head of Sina's news operations, Tang sensed he had a clear vision, one that was similar to that of Caixin Media. Her job would be to cover the 2016 elections that would sweep Donald Trump into the White House. 'They wanted to know what they called "the details of democracy",' she said. 'If I was reporting on a political rally, how did the police interact with the crowd? If it was a presser [press conference], what did I do to get called on for a question? Did I have to submit my question in advance? That kind of stuff.'

Her reports for Sina reached a much larger audience than her Caixin work, due in no small part to Sina owning the massive Weibo microblogging platform: 'They'd push your story out and millions of people would read it!' Tang had the responsibility of overseeing Sina's coverage of the US election campaigns and was hitting her stride in the summer of 2016, when the Republicans nominated Trump in Cleveland and the Democrats nominated Hillary Clinton in Philadelphia. Sina sent a big team to Cleveland, with Tang running the show. 'Conventions are fun, you get people from all over the country, and it's easy to talk to them and get their opinions,' she said. The fun didn't last long.

After the party conventions were already underway, a junior colleague told Tang that the call had been made back at Sina HQ in Beijing to 'shut it down'. All live and daily reporting would cease immediately, she was informed, but since they were already there, they should stay and get footage that could be used later. The decision blindsided Tang, who was not only disappointed, but confused as to why her boss preferred to tell her via a junior co-worker, when she was in charge there. It was an election far from China, and elections happen all over the world – what made this 'sensitive'?

The excitement of helping Chinese readers better understand the world was gone. 'I remember starting to feel very frustrated. Articles would be taken down for whatever reason, or words would be changed without anyone asking me. That was a big turning point.'

Despite her frustration, Tang did her best to be empathetic, but it was not easy. 'I didn't blame my colleagues, and I'm sure they were under a lot of pressure, but I began to feel like as a Taiwanese, I was not part of them,' she said, still sounding hurt years later. 'I felt them being careful with me, as if I might do something bad.'

Her colleagues were in fact under considerable pressure – from none other than Xi Jinping himself. Mere months before the GOP convention, Xi personally visited the operations of three of the Communist Party's most reliable mouthpieces: *The People's Daily*, *Xinhua News* and China Central Television (CCTV). Xi demanded unswerving loyalty not just from these party or government mouthpieces, but from *all* media in China.

'The party must be the media's surname,' as Xi put it, telling the country and the world that Chinese media would serve the needs of the party. It was reasonable for Chinese editors to be nervous about their jobs and personal safety, given that Xi had spent his first three-plus years at the top of the Communist Party aggressively purging his rivals and destroying their patronage networks – something unseen since the Mao years. Xi's message reached Tang's bosses in Beijing, who might have concluded that showing a competitive, democratic system – even one as flawed as the American model – might influence a Chinese audience into asking why Xi and the party offered no choice in how they are governed.

Tang persevered for a couple of years, during which she produced articles, breaking news reports and video segments day in and day out. Overall, life was good, even if sometimes colleagues would abruptly end WeChat messages for seemingly bizarre reasons.

In 2019, Tang travelled to North Carolina for some much-needed time off from the news grind. After waking up one morning, she saw her phone had been inundated with WeChat messages and missed phone calls. An unpleasant, sinking feeling set in as she learned that she had been accused of being a 'Taiwan independence supporter' by the well-known social media account of 'Guyanmuchan' (孤烟暮蝉).

Guyanmuchan, who still posts rabid nationalism on Chinese social media today, shared screenshots of Tang's social media posts cheering the passage of pro-Taiwan legislation by Congress, and her sharing of a tweet by Hong Kong democracy activist Joshua

Wong on Twitter (now X) to her millions of followers. The Chinese influencer's followers knew what to do, piling onto Tang's Twitter and Facebook accounts with insults and threats.

Rather than defend or console Tang, her editors and colleagues urged her to shut down her social media presence. Furthermore, her direct editor told her she would need to apologize to two of his higher-ups. She did so, but she described it as making her very uncomfortable. 'I genuinely felt bad,' she told me, even though she knew she had done nothing objectively wrong. 'There's a guilt about living in the free world – you don't know what Chinese people have to deal with.'

Tang trusted her team and didn't want colleagues to be fired on her account. Regardless, she realized she was done with her China experience. 'At the time, it was terrifying,' she said of the social media pile-on to which she was subjected by Guyanmuchan's followers. 'There was an invisible army of people trying to dredge up anything they could about me from my social media – I shouldn't have read their comments, but of course I did.'

One of the many accusations was that she was a spy, using her cultural fluency and good looks to get secrets for the highest bidder. 'The China I thought I knew suddenly felt like a boyfriend who was cheating on me,' Tang told me with a laugh. 'There were so many signs, how did I not see them!'

Tang estimates that since 2019, more than a thousand of her bylines have been scrubbed from the Chinese internet. She has fallen out of touch with most of her Chinese friends in Beijing, afraid to get them in trouble by messaging them, while they could feel uncomfortable contacting her – especially after she took a job at Radio Free Asia, which is blocked in China.

On an even more serious and unnerving note, the whole episode convinced her that Xi Jinping's China would not leave her or her country alone, probably for the rest of her life. 'I had the feeling that as a Taiwanese, I can't escape China, no matter what I do – this is the fate of Taiwanese people, of my generation, and it makes me very, very sad.'

Tang has no ill will towards China or the Chinese people, however, and told me she would like to go back and visit one day, under the right conditions. 'You know, you love and invest yourself so much in a place and envision a future there . . .' she said, trailing off. 'If it's safe, I would go back,' she said, pausing again, 'but I'm not sure it is.'

The American Media

Sadly, it is not just Chinese media that erases Taiwan's statehood. US media leads the way in the democratic world. Major outlets including *The New York Times*, *The Wall Street Journal* and *The Washington Post* all use awkward workarounds such as 'self-ruled island', 'democratic island', and 'self-ruled democracy' when referring to Taiwan, almost never using the space-saving and accurate term 'country'.[35]

Many foreign reporters based in Taiwan, myself included, consider the way most international publications dodge Taiwan's undeniable sovereignty indefensible. But for Taiwanese journalists working for international media who are forced to deny their country's statehood, this wilful denigration of their government is not just a professional issue, but also a personal one.

'As someone born and raised in Taiwan, it definitely sucks,' one Taiwanese journalist told me. 'But it's also part of being Taiwanese – you see Taiwan as its own country growing up, until you experience the wider world, and then you see its statehood denied.

'As a journalist, I would say there's definitely better ways to refer to Taiwan than fudges like "island" or unthinkingly calling it a "renegade province", which is an incorrect cliché prevalent in English writing and does not even exist in Chinese,' they told me.

'The country issue aside, foreign media don't give Taiwan a voice, instead emphasizing China's perspective on what Taiwan is, which is problematic as a journalist, because Taiwan has its own people and deserves a voice. Speaking as a journalist, this is simply unfair. You wouldn't do this to any other place in the world with such a one-sided narrative.'

Another Taiwanese journalist working for a major US media outlet expressed similar feelings of frustration. 'I definitely hate that I can't call Taiwan a country, but I'm also used to it. In sporting competitions, we have to call ourselves Chinese Taipei – we can't even use the ROC flag.'

Both reporters noted that things are gradually improving, with UK-based publications *The Economist* and *Financial Times* consistently referring to Taiwan as a country in their reporting, unlike their peers across the Atlantic. The reporters often successfully pushed back against their editors' insertion of 'mainland' to refer to China in a China–Taiwan context, given that it implies that Taiwan completes China – which is precisely why both the CCP and KMT use the term.

At the time of writing this book, there were no Asia-focused editors from major international media based in Taipei, which may partly explain why these ways of referring to Taiwan persist. As the people who assign and edit stories, editors are gatekeepers for Taiwanese stories written for a global audience. Taiwan's sovereignty and statehood are easier to ignore for editors based in Tokyo, Seoul, Singapore, DC or New York – all in countries that do not recognize Taiwan's government (but also do not recognize China's claim on Taiwan).

'Editors try to portray themselves as objective in not calling Taiwan a country, but I don't see it that way,' the second reporter told me. 'It's not objectivity, it's a decision. Legacy media outlets are afraid of shifting away from an editorial tradition that comes from reflecting China's way of seeing the Sinosphere and neglecting Taiwan's agency and how we see ourselves.

'There is a need to change this. People around the world need to know how Taiwanese see themselves – we need to address reality rather than caving to China's narrative,' they told me. 'Whether you call Taiwan a country or don't call it a country, someone benefits either way – either China or Taiwan – and the situation now is that the way we're reporting on Taiwan is benefitting China.'

The same journalist noted that they asked one of their editors about the policy of not calling Taiwan a country, with the editor replying that they thought Taiwan was indeed a country but, given that the American government doesn't recognize Taiwan as such, they follow Washington's lead.

I'd heard similar arguments from editors before, but it is difficult to not see this as outsourcing an internal style decision to a government, which would seem to contradict fundamental principles of objectivity in journalism. For example, the US does not have formal diplomatic relations with Pyongyang or Tehran, but there is nothing controversial about calling North Korea or Iran a country.

'I feel kind of pessimistic; even if I push for change it won't happen until there's an industry-wide change,' the first journalist told me. 'Only then will my organization change.' The second journalist laughed while recounting the struggle to get fair representation for Taiwan in American media stories about Taiwan, which they described as 'so annoying'.

'Most Taiwanese journalists working in foreign media that I know share the same goal: getting as many Taiwanese voices and stories out there as possible,' they told me. 'I understand that's very difficult for US media because it's US-centric, and the US looks at Taiwan through the lens of war.'

Then they paused.

'Sometimes I wonder if some people want a war.'

CHAPTER 11

The Indispensable Island

In early months of 2021, the world was one year into grappling with the continued impacts of the Covid-19 pandemic that had emerged from China. Aside from the human toll and the economic damage the pandemic had unleashed, it also brought to light a critical dependency that caught many developed nations and their biggest corporations by surprise – the importance of Taiwan's semiconductor supply industry. Among the hardest hit were automakers, which rely on approximately 1,400 chips per vehicle. The pandemic-induced closure of auto production lines in 2020 led chipmakers to divert production to consumer electronics such as laptops and tablets, which saw increased demand during lockdowns as much of the world's office workforce began working from home. This pivot left the automotive sector in a bind when production resumed, as there weren't enough chips to go around. Suddenly, the world woke up to Taiwan's technological importance.

For the uninitiated, semiconductors are small, flat chips of silicon that contain electric circuits and are essential to all electronic devices. Today, Taiwan is responsible for more than 60 per cent of global semiconductor production.[1] Most of these chips are made by one company: the Taiwan Semiconductor Manufacturing Company. What makes TSMC even more important is the fact that it supplies more than 90 per cent of the world's most advanced chips, which are crucial to everything from mobile phones to weapons systems.

Before the chip shortage crisis triggered by Covid-19, TSMC and the broader Taiwanese semiconductor sector were not widely understood beyond the worlds of tech or financial markets. However, the pandemic swiftly altered this. For governments and boardrooms around the world, learning that their economies or companies were so dependent on Taiwan, and quite often TSMC specifically, was a wake-up call. Since that inflection point, politicians and other decision-makers who had previously paid little attention to Taiwan have had no choice but to consider its outsized importance to the tech supply chain that is integrated into virtually every corner of the global economy.

It was a particularly startling realization for Germany's \$300 billion automotive sector. The German government – which, under then-chancellor Angela Merkel, had previously shown little interest in Taiwan while prioritizing relations with Beijing – suddenly realized it needed to talk more with Taiwan. German industry was in urgent need of the chips that were crucial in vehicle manufacture. In a letter that may have elicited some schadenfreude in Taiwan, German economic minister Peter Altmaier asked his Taiwanese counterpart, Wang Mei-hua, if she could ask TSMC to make sure to keep the chips flowing to German automakers. 'I would be pleased if you could take on this matter and underline the importance of additional semiconductor capacities for the German automotive industry to TSMC,' Altmaier wrote.[2]

The chips did flow, but the richness of the request by a German government that has historically kept its distance from democratic Taiwan did not go unnoticed in Taipei. Two years later in 2023, as Berlin pushed for TSMC to commit to a factory in Germany, Taiwan foreign minister Joseph Wu told European media that his country was happy to help its friends however possible, but that a closer bilateral relationship would require a new way of looking at

Taiwan. He noted that the US and Japan were 'like-minded countries' – Taiwan diplo-speak for close unofficial allies – and both had TSMC facilities under construction.³

'We certainly hope that other countries who want to attract TSMC to make investment can also think about the situation Taiwan is in, or TSMC's position in Taiwan, and the position Taiwan is seeing in this geo-strategical landscape,' Wu said.⁴

His message was clear: you get more from us more than we do from you, yet you barely acknowledge our existence beyond our role as your tech supplier, at a time when we are facing an existential threat from your preferred trade partner. It was an uncharacteristically pithy statement from a Taiwanese foreign minister, but times had changed, and everyone knew it. For years, many developed democracies had handled their relationships with Taiwan cautiously, wary of upsetting China. Suddenly, Taiwan had huge diplomatic leverage.

In September 2024, the German frigate FGS *Baden-Wurttemberg* and logistic ship FGS *Frankfurt am Main* passed through the Taiwan Strait. Shadowed by Chinese vessels, the German ships were exercising what is known as 'freedom of navigation', highlighting that unlike Beijing, Berlin considered the Taiwan Strait to be international waters. It was the first time that German naval vessels had transited the strait in two decades.⁵ Less than four weeks earlier, the groundbreaking for the first TSMC chipmaking facility in Europe took place in Dresden, Germany. Among those present for the ceremony were German Chancellor Olaf Scholz, TSMC Chief Executive Officer C. C. Wei and European Commission President Ursula von der Leyen. The heads of German firms Infineon Technologies, NXP Semiconductors and Robert Bosch were also in attendance. Each of the companies owned a 10 per cent share in the €10 billion (US$11 billion) venture, with TSMC holding the remaining 70 per cent. With roughly half of the project funded by German government subsidies, this investment was a strategic move to bolster Germany's and the broader EU's semiconductor autonomy, aligning with the EU's goal to produce 20 per cent of the world's semiconductors by 2030. To achieve such a goal, it had little choice but to turn to TSMC, the world's largest contract chipmaker.[6]

Neither Taiwan nor TSMC were unfamiliar with being leaned on for their ever-important semiconductor technology. In 2020, the Trump administration essentially strong-armed both into a commitment to invest in Phoenix, Arizona. It was politically motivated on both sides. For the US, it would reduce the feeling of being heavily reliant upon a special friend that it hoped it wouldn't need to defend. For the Tsai administration, it was an acknowledgement that Taiwan was too important to ignore any longer. In April 2024, TSMC's first Arizona 'fab', as chip plants are known in the industry, began trial production of chips at the 4-nanometer level. While the chips are not quite at the level of sophistication of the most cutting-edge semiconductors it produces in Taiwan – it is currently testing production at the 1.4-nanometer level – it is far ahead of the most advanced chips produced in China. After initial delays due to a lack of qualified staff and friction with American staff over the company's demanding corporate culture – not normally an issue in Taiwan – TSMC's Wei expressed optimism for the Arizona project.

'We now expect volume production of our first fab to start in the beginning of 2025, and are confident to deliver the same level of manufacturing quality and reliability from our fab in Arizona as from our fabs in Taiwan,' Wei said in an investor call in October 2024. Around the same time, the president of TSMC's US division, Rick Cassidy, told reporters in a webinar that the Arizona fab's yield – the measure of usable chips – was roughly four percentage points higher than comparable TSMC facilities in Taiwan.[7]

This was not only good news for TSMC, but also for Donald

Trump, who would be elected as US President for the second time less than a month later. It was the Trump administration, after all, that had both pushed Taiwan's government and incentivized the Taiwanese chipmaker to set up shop in Arizona, where a small tech supply chain already existed. In exchange for building three football stadium-sized fabs outside Phoenix, TSMC stands to receive US$6.6 billion in US government grants, as well as US$5 billion in loans and tax credits of 25 per cent.[8]

Across the Pacific, far from the Arizona desert, TSMC was also establishing a presence in Japan. Kumamoto, a sleepy prefecture on the southern island of Kyushu, is known for being home to one of the country's three most-renowned castles. Originally built on a hilltop in the fifteenth-century, Kumamoto Castle was gradually expanded, adding a towering keep to its curved stone walls and wooden overhangs. It famously survived being besieged and attacked with nineteenth-century weaponry during the Satsuma Rebellion of 1877. Now Kumamoto's most noteworthy building is a state-of-the-art TSMC chip fab. In February 2024, TSMC launched production at the first of up to three planned fabs in Kumamoto, with construction of the second commencing at the end of the year. As in Phoenix and Dresden, TSMC's investment in Kumamoto is receiving generous government incentives – while also driving TSMC suppliers and other Taiwanese companies to invest in the area. Regional lender Kyushu Financial Group estimates that in the coming decade, TSMC and other companies in the local industrial ecosystem will add US$80 billion to the economy of Kumamoto, which is home to around 750,000 people.[9]

The public discourse regarding TSMC's global expansion, as with almost everything else in Taiwan, has been contentious. The KMT argues that the US's encouragement of TSMC's overseas ventures is motivated by a desire to relocate TSMC's most advanced technology onto its own soil so that it no longer needs to defend Taiwan. This perception is bolstered by comments by some

American politicians and aspiring Trump administration members. On the campaign trail for his failed 2023 bid for the Republican presidential nomination, Vivek Ramaswamy suggested ditching the policy of strategic ambiguity – not saying whether the US would defend Taiwan in the event of a Chinese attack – for strategic clarity. At least for a while, that is.

'I'm clear, we will defend Taiwan, at least until we have achieved semiconductor independence in this country, at which point we will re-evaluate,' Ramaswamy said.[10] After being criticized by many of his fellow Republicans for his comment, Ramaswamy appeared to conflate Washington's one-China *policy* with Beijing's One China *Principle*. 'Neither party even recognizes Taiwan as a nation right now,' he said in an interview. 'Both parties embrace the one-China policy.'

Semiconductor independence might sound good to America-firsters but it is a pipe dream, at least anytime in the near future. Again, as early as 2022, TSMC chairman Mark Liu has said that an invasion by China would effectively end TSMC's operations.[11] This would presumably include TSMC facilities overseas, such as the Phoenix plant. The Phoenix facility, despite being on American soil, would likely face operational disruptions due to losing its headquarters and the intricate, immensely sophisticated supply chain rooted in Taiwan. TSMC has also vowed to keep its most advanced chip production and research and development in Taiwan.

Semiconductor Industry Dominance

It is difficult to wrap one's brain around the processes TSMC has deployed to achieve industry dominance, or the costs it incurs to maintain its industry leadership.

The first integrated circuits were developed in the middle of the twentieth century. They replaced the bulky and inefficient vacuum tubes used in the earliest computers and other electronic devices, but only had a handful of transistors – which register the individual ones and zeroes of the binary language used by computers – per chip. Today, TSMC's cutting-edge chips have billions of transistors etched onto them, a testament to the monumental strides made in semiconductor technology, in which the Taiwanese company is the undisputed global manufacturing leader.

The method TSMC employs to achieve such feats of engineering, extreme ultraviolet lithography (EUV), sounds more like black magic than science. The EUV process employs a series of highly sophisticated steps to create the right light necessary for etching the minuscule transistors of a semiconductor chip. At the core of this process are droplets of tin plasma, roughly 25 microns (millionths of a metre) in diameter, which are emitted from a generator at a speed of 70 metres per second. While they fall, the plasma droplets are initially shot with a low-intensity laser pulse that flattens them, before a more powerful laser pulse vaporizes the droplets, creating a plasma that emits EUV light. In order to produce enough light to etch semiconductors, this process is repeated 50,000 times *per second*.[12]

In 1965, Gordon Moore, who co-founded Fairchild Semiconductor, one of the first major American chipmakers, and, later, Intel, made an observation. Technological advances were enabling the number of transistors that could be crammed onto one integrated circuit to double every year. By 1975, he refined his prediction, anticipating a doubling every two years. This prediction, known as Moore's law, was only intended to be accurate for the following decade, but it has held true for half a century now. It influences how companies like TSMC strategize for the future – it is a benchmark for long-term planning, as well as for setting research and development goals. It is not an exaggeration to say that Moore's law has driven the technological changes, productivity gains and economic development of the past fifty years. As TSMC chases Moore's law for clients such as Apple and Nvidia, it burns through massive amounts of money on the facilities and research and development that are required for this quest. In TSMC's third quarter earnings call for 2024, Senior Vice President and Chief Financial Officer Wendell Huang estimated the company's capital expenditures for the year at a little over US$30 billion.[13] For comparison, Taiwan's entire defence budget for the same year was two-thirds of that number, at just over US$20 billion.

Taiwan is prone to earthquakes, and TSMC says that its facilities can withstand tremors of up to 9.0 in magnitude on the Richter scale. They are not, however, war-proof. This means that the Sword of Damocles that hangs over Taiwan hangs over all the world's major economies – including China's. Ken Griffin, founder and CEO of the hedge fund Citadel, told CNBC in early 2024 that if there were a 'rupture around Taiwan it would be catastrophic to both the Chinese and to the American economy. And by catastrophic I think you're looking at Great Depression circumstances.'[14]

After being pressed, Griffin expounded on the potential consequences.

'If we lost access to Taiwanese semiconductors,' he asked, 'how many weeks until Tesla stops making cars? Or GM? Or Ford? Or Boeing stops making planes? Those chips are used in every part of our economy. Estimates range from a GDP hit of between 8 and 10 per cent if we lost access to Taiwanese semiconductors. So, it's really important as a matter of national economic security that we're able to maintain peace in that region of the world.'

In today's hyper-globalized world, nearly every country, from basic commodities exporters up to the most developed economies, would experience an economic pain that would make the recent pandemic seem mild. And that's before considering that an elective war of conquest by Xi Jinping could also end up involving the US, Japan, the Philippines and even Australia.

Given the non-transparent nature of his rule in China, there is no way to accurately gauge Xi's appetite for risk with regards to Taiwan. China's recent economic slowdown has presented Xi with challenges he didn't have to worry about before, and an attempt to invade Taiwan would compound problems for the Chinese economy. Yet CCP leaders have historically been willing to use the CCP's armed wing, the PLA, in order to ensure the party remains in control of China. Notable examples of this include the Korean War, the Cultural Revolution and the nationwide crackdown on democracy protesters in June 1989. All incurred large human costs on China's side, and all kept the CCP in power. Those who argue that China wouldn't invade Taiwan because it would be economically catastrophic for China need only look back a few years to those who said Beijing would never crack down on the financial golden goose that was Hong Kong.

The CCP's historic willingness to impose suffering on the Chinese people is but one reason not to discount a PLA attempt to seize Taiwan. Another is that China has made major gains in manufacturing the semiconductors a few rungs down from TSMC's most advanced chips. Were TSMC to no longer exist, China would suddenly become a global chip power – at a high cost, no doubt, but in the long term, it could be an appealing result: countries seeking to rebuild their economies post-depression would likely have to do so on Beijing's terms.

To summarize, the world's most important technology is produced at a near-monopoly level by one company located in Taiwan, a country that most of the world does not officially recognize and which is threatened by the world's largest military.

So, how did we get here?

The first step in Taiwan's decades-long journey from isolated military dictatorship to the centre of global technology was a business trip to Taipei by two Texas Instruments executives in 1968. Their meeting with K. T. Li, the ROC's economic minister, laid the groundwork for a seismic shift in the global technological landscape.

Dallas native Mark Shepherd was the head of semiconductors at Texas Instruments, then one of the world's top chip manufacturers. As a boy, he displayed an intense interest in electronics that would define the rest of his life. He built his first vacuum tube at the age of six and his first radio at the age of seven. Shepherd was one of the four engineers that Texas Instruments sent to Bell Laboratories in 1952 to learn about the company's transistor, which it had invented in 1947. He later witnessed the 1958 demonstration of one of the first integrated circuits by his colleague, Jack Kilby. When Shepherd visited Taiwan in 1968 at the age of forty-five, he was already a semiconductor industry veteran. He was joined on his trip by a junior colleague – a promising thirty-seven-year-old engineer named Morris Chang.

Born in China's Zhejiang province, Chang's family had fled Mao's Communist revolution, ultimately settling in the US. Chang briefly attended Harvard before transferring to MIT, where he switched from studying the humanities to mechanical engineering. After graduation, Chang took a job with Sylvania, where he worked on their integrated circuits. After joining Texas Instruments in 1958, Chang made notable advancements, quickly ascending through the company's ranks. The pipe-smoking Chang earned a reputation for genius and his scathing criticism of underlings, and, three years later in 1961, he became manager of a large engineering section. By 1968, as he accompanied Shepherd to Taiwan, Chang had risen to general manager of Texas Instruments' most important semiconductor department – integrated circuits – and was managing more than 2,000 people.[15]

K. T. Li was economic minister despite never having studied economics – he studied nuclear physics at Cambridge. Li embraced a hands-on, practical economic strategy, and would later be hailed as the 'father of Taiwan's economic miracle'. Having been a crucial figure in the transformation of Taiwan from an agrarian economy to a tech powerhouse, the former KMT official is held in high esteem on both sides of Taiwan's political spectrum. His steering of the economy was marked by careful planning and an emphasis on stable development. While the miracle was not the result of any one individual, Li's cautious approach facilitated Taiwan's rise as an international trade and tech power.

Born in Nanjing in 1910, Li had escaped to Taiwan with the KMT in 1948. Having served for a decade in the KMT government prior to arriving in Taiwan, Li was wary of the threat from the CCP and hyperinflation: both had been major contributing factors to the KMT's loss of China. As a result of these concerns, Li worked to ensure that Taiwan was not just politically but economically important to the US (an approach that would be revived decades later under the presidency of Tsai Ing-wen).

In the late 1960s, the global semiconductor industry began offshoring labour to Asia, leveraging the region's cost efficiencies, high productivity and absence of unionized labour. Taiwan, with its skilled yet affordable workforce, offered an ideal environment for manufacturing. With Shepherd and Chang's encouragement, Texas Instruments decided to take a gamble on Taiwan in 1968. This move marked the beginning of Taiwan's journey towards a central role in the global semiconductor manufacturing industry. Production at its first facility began in 1969. Eleven years later, in 1980, it had produced its billionth semiconductor.[16]

The 1970s saw Silicon Valley fall behind Japan in semiconductor manufacturing. However, the 1980s brought a resurgent US tech sector, partly fuelled by increased investment in facilities in Taiwan. This shift occurred amid significant changing geopolitical realities, notably Mao Zedong's death in 1976 and China's subsequent economic opening under Deng Xiaoping. These changes presented both opportunities and challenges for Taiwan's tech sector, which, despite initial setbacks, leveraged these circumstances to its advantage.

K. T. Li was no doubt unnerved by Washington's derecognition of the ROC in 1979, which was not only driven by the political goal of containing the Soviet Union, but also the economic objective of entering and making money in the world's most populous country. Nonetheless, Taiwan embarked on an ambitious journey to reshape its role in the global economy. By the middle of the 1980s, the Taiwan economic miracle was well underway. Between 1970 and 1985, Taiwan's GDP shot up nearly 1,000 per cent, from US$5.8 billion to US$63.6 billion. Per capita GDP jumped 735 per cent, from $397 to $3,314. This economic leap was not just a testament to Taiwan's capacity for rapid growth but also a strategic pivot from a reliance on low-value manufacturing towards becoming a global technology and innovation hub.

Taiwan was initially a major player in the production of low-value items like footwear, umbrellas and plastic toys. Chiang Ching-kuo and his economic mandarins knew that if Taiwan's economy remained cantered on such low-value items, it was doomed to be hollowed out by China and other countries where labour and land costs were lower. Still a military dictatorship, the KMT used sustained economic growth, rising incomes and improved health and education to claim legitimacy. Maintaining Taiwan's upward economic trajectory was a political imperative. A critical element of this transition was the emphasis on science and technology. This included a concerted effort to send Taiwanese students to learn from top universities in the US and elsewhere to cultivate a highly skilled workforce.

Two pivotal decisions further laid the groundwork for Taiwan's tech emergence. In 1973, it founded the Industrial Technology Research Institute (ITRI), which would drive technological development and eventually serve as an incubator for tech-focused companies. Then, seven years later in 1980, the government would establish Hsinchu Science Park. The latter had a ten-year, US$500 million fund to turn 600 hectares into Taiwan's answer to Silicon Valley. The goals of the park were threefold: to inject new energy into the economy, establish a homegrown high-technology base and address a growing brain-drain problem.

In 1980, the first Taiwanese semiconductor company, United Microelectronics Corporation (UMC), was founded in Hsinchu Science Park. The company was a spin-off of ITRI, which designed and produced its own chips. Taiwan was finally in the semiconductor race, but it was at the rear of the pack. UMC had been built upon technology that ITRI had licensed from the American company RCA in 1975 – technology that was by then already a generation behind. Taiwan's semiconductor aspirations required more than money, engineers and outdated technology. Someone with vision, leadership skills and experience was needed. Someone like Morris Chang.

Morris Chang had made a major impact at Texas Instruments but had been overlooked for the role of CEO, and found himself at a crossroads. Despite his significant contributions, he faced career stagnation. 'I was still a senior vice president but effectively I felt that I had been put out to the pasture,' Chang said in a 2007 interview.[17] So in 1983, he left. Taiwan came calling two years later. They offered him the job of President of ITRI. He took it. Chang was fifty-four years old and had plenty of energy and ambition left. He wanted a new kind of career, and he got one.

'Rather than being a corporate executive, the ITRI job, the presidency of ITRI would mean serving the whole industry, serving the whole Taiwan industry,' Chang said. 'This was a completely new kind of a job for me, and also the environment would be completely different. A Taiwanese environment as opposed to the US environment. The kind of people that I would work with would also be very different. So it was the newness that appealed to me, the difference from what I had been doing for several decades.'[18]

'You have a lot of experience managing large semiconductor businesses, so you seem to be ideal to start a new company in Taiwan in semiconductors,' Li told Chang upon his arrival at his new job. 'You come back, tell me and tell the premier how much money you need.'

He was given less than a week to formulate his vision for a new semiconductor company and present it to the government. The presentation he gave formed the basis of the company that Chang would found with government backing two years later: TSMC. Drawing inspiration from the seminal 1978 textbook by the researchers Carver Mead and Lynn Conway – *Introduction to VLSI Systems* – Chang envisaged a novel business model in the semiconductor industry. While not specifically advocating the separation of semiconductor design and manufacturing – all semiconductor companies did both at the time – Chang's takeaway from the book was that it was possible. This idea would revolutionize the semiconductor industry.

Chang knew from experience how cutthroat the semiconductor industry was, and recognized the immense challenge he was facing in trying to make a competitive company from scratch in Taiwan. When he looked at what resources Taiwan had to push towards that lofty goal, he saw a country that was hopelessly weak in nearly every category that mattered: research and development, circuit design, integrated circuit product design, sales and marketing, and intellectual property. The only area where Taiwan had something – anything – to offer the project was that it was already manufacturing semiconductors, which were nowhere near the cutting edge. The only possible way that Taiwan's new semiconductor company had a chance of surviving was to create what is known as a pure-play foundry, a fab that only made chips for other companies and had no products of its own. Given Taiwan's weakness in everything else, he saw no other option. The government decided to back his idea. It was less well-received by his peers in the semiconductor industry as, at the time, there were basically no 'fabless' companies – firms who only designed chips.

'Well, people just dismissed it, you know,' he recalled. '"What the hell is Taiwan doing? What the hell is Morris Chang doing?" They really didn't think that it was going to go anywhere. There was no market because there was very little fabless industry, almost none. No fabless industry. So who are you going to sell these wafers to?'

Chang's idea was that TSMC would initially help soak up excess production demand for companies such as Intel, Texas Instruments and Motorola, and that fabless companies would come along later. Both bets were proven right, with TSMC only losing money in 1987 (its first year) and in 1991. At the time of the writing of this book, TSMC was the world's tenth most valuable company, with a market capitalization of more than US$950 billion. Morris Chang was valued at US$4.2 billion. When Chang founded TSMC with money from the government, the Dutch electronics company Philips and individual Taiwanese investors – many of whom were strongarmed by ROC officials into supporting the start-up – Taiwan was at the bottom of the semiconductor food pyramid, but forty years later, it would be the apex predator.

Chang's vision fundamentally transformed the semiconductor industry landscape, not only for Taiwan but for the global market. It unleashed a wave of fabless semiconductor companies, who only had to invest in designing chips, saving them the growing cost and headache of producing them themselves. TSMC effectively lowered the cost barrier to getting into the chip game and spurred innovation, leading to the founding of Taiwanese fabless semiconductor companies such as MediaTek. Another ITRI spin-off itself, MediaTek capitalized on the opportunities provided by TSMC's manufacturing capabilities to become a leader in designing chips for mobile phones.

By the end of the 1990s, an increasingly ambitious Chang had set his sights on overcoming the industry leader, Intel. Over the next decade, TSMC gradually closed the gap with its American rival. Chang stepped down as company CEO, while staying on as chairman. That would change after the global financial crisis of

2008 blindsided the semiconductor industry, with most companies downsizing to reduce costs. Chang's successor as CEO, Rick Tsai, did the same. It was the first time the company had laid off employees. Undeterred by age, Chang, at seventy-seven years old, made a dramatic return as CEO, revoking the layoffs and significantly increasing TSMC's investment. Investors may not have been pleased, but the move further solidified TSMC's position in the industry.

A pivotal moment came in 2010 when Apple, seeking alternatives to Samsung for manufacturing chips for the iPhone, initiated discussions with TSMC. After on-and-off negotiations over months, Apple went silent. It was talking with Intel, who ended up turning down the chance to make chips for the iPhone because it didn't like Apple's terms. Chang ended up landing the contract for TSMC, after which he borrowed US$7 billion to add necessary capacity. Despite Apple's tough terms, his gut told him that the scale of the deal would get him closer to his goal of overtaking Intel. Chang's instincts were once again proven correct – Apple now accounts for one-fifth of TSMC revenue.

Chang would retire again in 2018, confident in TSMC's technological lead. Having been involved in the semiconductor industry since its earliest days, and having outlived all the original players in the game, Chang has a level of experience and wisdom that decision-makers at other companies simply couldn't match. His vision and success with TSMC not only led to the founding of numerous Taiwanese fabless semiconductor companies, but also to an ecosystem of Taiwanese companies serving the chip giant's complex and growing needs.

At the Forefront of AI

Unlike Intel, which missed out on the mobile revolution, TSMC appears poised to continue its development through the next major technological shift: artificial intelligence (AI). TSMC produces chips for the world's leading AI chip design company: Nvidia. Under the leadership of CEO and co-founder Jensen Huang, Nvidia has surged to remarkable heights, with a market capitalization of nearly US$2.5 trillion at the time of writing, ranking as the third most valuable company globally after Apple and Microsoft.

Jensen Huang was born in Tainan in 1961, with his family moving to the US when he was a child. In 1993, after obtaining a master's degree in electrical engineering from Stanford, thirty-year-old Huang and two partners decided to found Nvidia during a meeting at a Denny's restaurant in San Jose. Huang's vision for the company was clear from the outset. Noticing an opportunity in the gaming industry's graphics chips – a sector overlooked by behemoths like Intel and Samsung – Huang reached out to Morris Chang of TSMC. After a phone call with Huang, Chang decided to take Nvidia on as a client. 'I liked him,' Chang said in a 2023 interview.[19]

The genius of Nvidia's strategy lay in harnessing the parallel processing capabilities of graphics processing units (GPUs) for general computing purposes, beyond their traditional role in video gaming. In 2023, the AI boom led to Nvidia becoming a trillion-dollar company, driven by its position as a leading producer of data-centre chips with AI capabilities.

With an estimated wealth of more than US$89 billion, the sixty-two-year-old Jensen Huang is the seventeenth-wealthiest person on the planet. Yet he has never forgotten his Taiwanese roots. On his frequent trips to Taiwan, the billionaire is often spotted at night markets, where he pays between $1 and $4 for local snacks such as cherry tomatoes in ginger sauce, oyster omelettes or stuffed squid, all while taking selfies with admirers and stall workers. In early 2024, he took Morris Chang to a night market – it was Chang's first time visiting one. Around that time, he also gave a keynote speech about the new industrial shift he believes will be driven by AI. Donning his signature black leather jacket and backed by the logos of forty-three of Nvidia's key partners and suppliers, Huang's message was clear: Taiwan is poised to play a pivotal role in the AI revolution. The silver-haired Huang addressed the packed auditorium for two hours, urging Taiwan to aim for dominance in AI.

It goes without saying that Jensen Huang is immensely popular in Taiwan, where his local boy from Tainan persona comes across as pitch-perfect and genuine. In addition to speaking English and Mandarin in public appearances, he often switches to Taiwanese. He is clearly grateful to Morris Chang and TSMC, as well as to Taiwan. 'Without Morris and TSMC, there would be no Nvidia,' Huang said in a 2023 interview.[20] Speaking in San Jose at a 2024 banquet for Nvidia's Taiwanese supply chain partners, he called himself 'a very good ambassador of Taiwan'; as he understands that it is at the centre of the AI revolution. 'We must make sure we tell that story this time,' he said, before raising a glass of wine to the audience and saying, in Mandarin: 'Go Taiwan!'[21]

Taiwan's evolution into a global technology powerhouse, particularly in the realm of semiconductors, is a remarkable journey of innovation and forward thinking. The island country has meticulously carved out a niche that is now foundational to the world's digital economy, with ramifications for everything from consumer products to weapons systems. At the heart of this transformation are entities like TSMC and visionaries like Morris Chang and Jensen Huang, whose contributions have elevated Taiwan onto the global stage and underscored the strategic significance of its technological prowess. This position is pivotal at a time when global tech leaders are vying for supremacy in AI – a field that promises to redefine industries, economies and societal norms.

EPILOGUE
Eye of the Typhoon

At the outset of 2025 – or Year 114 of the Republic of China calendar – it was clear in Taiwan that the geopolitical headwinds faced by President Lai Ching-te are only going to get stronger. At risk are Taiwan's economy, diplomacy and sovereignty itself. The main drivers of growing uncertainty regarding Taiwan's future are a familiar trio: the CCP, the US, and the Chinese Nationalist Party (also known as the KMT).

Shortly after the Chinese navy launched its first Type 076 amphibious assault carrier – christened the *Sichuan* – state broadcaster CCTV confirmed that it would be the world's first drone carrier. The gigantic *Sichuan* is heavier than France's largest aircraft carrier, and features a full-length runway and electromagnetic catapult facilitating the launch of larger aircraft. In an invasion scenario, the *Sichuan* would deploy a wide arsenal of military assets, potentially including the GJ-11 stealth drone. Dubbed 'Sharp Sword', the aircraft is capable of deploying swarming airborne decoys, electronic warfare systems and precision guided munitions.[1] The *Sichuan* may be ready for service as soon as the end of 2026, Chinese media has reported. Among the crewed and uncrewed airborne and amphibious systems it is expected to deploy are hovercrafts that would ferry armed robotic dogs to land, where they could assist in operations in myriad ways. At more than 250 metres, the length of the *Sichuan* is comparable to a fifty-storey building laid on its side. At the Hudong shipyard in Shanghai, Chinese naval officers and sailors stood in the January cold for the *Sichuan*'s christening, applauding after a bottle of champagne smashed against the vessel's side, and explosions of coloured smoke were launched high up into the air along its length.

Further down China's coast in Guangzhou, another Chinese shipbuilder was hurrying construction of a fleet of unusual barges carrying 120-metre-long extendable bridges. Western defence analysts concluded the barges were being purpose-built for a Taiwan invasion.[2] The barges under construction, which may number as many as seven, appear similar to the two temporary harbours designed by the British admiralty for the Allied invasion of Normandy – an operation that would be dwarfed by a Chinese amphibious landing on Taiwan. In battle, the barges would serve as piers, enabling the unloading of tanks, trucks and troops brought over on giant cargo ships or civilian ferries. The barges further complicate matters for Taiwanese defence planners by expanding the number of potential landing sites from which China can choose, which had previously been limited to a small number of ports and beaches.

Many analysts who follow China's military build-up see the rapid construction of novel vessels designed for a Taiwan invasion as indicative that Beijing is nearing what it considers readiness to go to war. 'I honestly cannot think of too many developments that could be more of a red flag for Taiwanese and US/allied defense planners that the PLA is making real its direction from Xi Jinping to have the capability to invade Taiwan by 2027,' defence analyst and consultant Tom Shugart said of the barges on the social media site X.[3]

'This does not mean, of course, that the PRC/CCP will actually decide to try such a risky course of action,' Shugart noted, 'but even lower-level options like a quarantine/blockade/bombardment may have far more impact if they also have a credible threat of invasion behind them.'

As if that isn't enough, China also appears to be receiving help with its campaign of intimidation against Taiwan from its biggest friend and ally: Russia. Data contained within a Japanese government report released in January 2025 suggested that Russian warplanes were flying just outside Japanese airspace so that China could keep up its daily harassment of Taiwan. The report noted that scrambles by Japan's Air Self-Defence Force in the first three quarters of fiscal year 2024 saw a dramatic increase in the number of Russian jets intercepted. At the same time, sorties aimed at intercepting Chinese warplanes near Japanese airspace dropped.[4]

Some analysts interpreted the spike in Russian buzzing of Japanese air space as helping pick up slack so that China can maintain its campaign to gradually get closer to and more familiar with Taiwanese territory, while also wearing down Taiwan's much-smaller air force. Vladimir Putin is, of course, deeply indebted to his indispensable ally Xi Jinping for his support in Russia's war in Ukraine. Putin is a vocal supporter of China's annexation of Taiwan – Russia's assistance in a Chinese war of conquest against Taiwan is highly likely given the debt he owes Xi.

The return of Donald Trump to the White House on 20 January 2025 was attended by Taiwanese delegations from the DPP, KMT and Legislative Yuan. While Trump's first administration is generally considered the most Taiwan-friendly since the US ended recognition of the ROC government in 1979, it is unclear how he will manage relations with Taipei and Beijing the second time around.

'We reject any effort to coerce, intimidate, and/or forcibly drive Taiwan to do whatever China wants them to do,' said Marco Rubio in a Senate hearing prior to his unanimous congressional approval as Secretary of State.[5] Describing the defence of Taiwan from China as 'critical' for the US, Rubio warned Beijing that it needs to 'stop messing around with Taiwan'.

Such words may have provided solace for Taiwanese concerned that US support under a second Trump administration might wane or be less predictable compared to before. Such sentiments were soon challenged, however. In his first week back on the job, Trump ordered Rubio to issue a 'stop-work notice' for all existing foreign aid, including military financing to Taiwan, while pausing new aid until he had reviewed it.[6] A month later, Rubio exempted $870 million in defence aid to Taiwan and $336 billion for the Philippines, signalling continued support for US interests in the region.[7] Any reassurance felt by Taiwanese officials was likely undercut by the extraordinarily undiplomatic treatment of Ukraine President Volodymyr Zelenskyy by Trump and Vice President JD Vance only days later in the White House. Rubio's silence at the scene likely diminished many Taiwanese people's hope that this old friend of Taiwan (and Ukraine) might be able to rein in possible desires by Trump and his special adviser, Elon Musk, to engage in the kind of horse-trading with Xi over Taiwan that Roosevelt and Chiang had done eight decades earlier.

Before the shock of his abandonment of Ukraine had worn off, Trump was back in front of reporters at the White House days later with TSMC CEO C. C. Wei to announce the biggest-ever foreign investment in the US. The Taiwanese tech behemoth had agreed to invest $100 million in the construction of five additional facilities in Arizona. TSMC's investment will build three chip fabs, two advanced packaging facilities and one research and development centre, Wei said.[8]

'This is a tremendous move by the most powerful company in the world,' the American president said. Later in the press conference, a reporter asked Trump whether a greater presence by TSMC on American soil could mitigate the damage to the US should China attack Taiwan.

'That would be a catastrophic event, obviously,' Trump said. 'But it will at least give us a position where we have – in this very, very important business, we would have a very big part of it in the United States. So, it would have a big impact if something should happen with Taiwan.' Several reporters continued to ask Trump questions about Ukraine. None asked him if TSMC's investment reaffirmed the American commitment to Taiwan.[9]

A year before Trump's return, in the week following Lai Ching-te's January 2024 election, I met with Congressman Ami Bera, a representative from California who co-chairs the large and influential Taiwan caucus. The California Democrat had been joined on the trip by his fellow co-chair Mario Díaz-Balart, a Republican from Florida.

'We thought it was important to send a message from the Congressional perspective that the US–Taiwan relationship is as strong as ever,' Bera told me at a hotel meeting room in Taipei's old Japanese quarter. 'The United States is not interested in changing the status quo, but the level of aggression, the level of grey-zone tactics, across the strait but also in the region, that Beijing is engaged in, forces us to have to react and confront that,' he added. 'You don't have to look very far to see how China has changed the norm in the South China Sea. For that to happen in the Taiwan Strait, where there is so much movement of goods and services – we can't let that happen.'

When I asked him how he thought Congress would handle Taiwan under a second Trump administration, Bera told me the historical bipartisan commitment of Congress to Taiwan would be 'as strong as ever'. 'Just the number of bills, the number of resolutions, the number of statements that are coming out in support of Taiwan are greater than anything I've seen in my twelve years in Congress. You're seeing resources supported in a way that we probably wouldn't have seen even five years ago.'

Later in 2024, I spoke with Ivan Kanapathy, one of the most

informed Americans on Taiwan's military situation. Kanapathy served on the White House's national security staff as director for China, Taiwan and Mongolia, and deputy senior director for Asian affairs under both Trump and Biden from 2018 through 2021. Prior to that, the retired Marine lieutenant colonel was military attaché at the American Institute in Taiwan from 2014 to 2017, giving him deep access to Taiwan's military personnel and facilities. Almost immediately after taking office on 20 January 2025, Trump named Kanapathy his senior director for Asia at the National Security Council. When I spoke with Kanapathy prior to his appointment, I asked him what he thought the ramifications of a successful Chinese invasion would be.

'The loss of Taiwan would mean the end of US credibility in the region and the world – and effectively the end of the US-led post-WWII order,' Kanapathy told me. 'In Asia, Japan and South Korea – and perhaps Australia – would develop nuclear weapons. Southeast Asian countries might do the same. All countries would be less resistant to China's coercion and bullying, effectively surrendering the East and South China Seas to PRC control.'

With that in mind, I asked him what he saw as the most pressing issue for Taiwan's defence. 'The most important thing Taiwan can do is increase – by tenfold or more – the number of mobile, ground-based anti-air munitions, such as Stingers, on the island,' he said. 'Effectively denying air superiority is critical to preventing a PLA takeover.'

It remains to be seen how Trump will handle arms sales to Taiwan, which in late 2024 were suffering from a delivery backlog of US$20 billion. A statement from the Washington-based US–Taiwan Business Council issued weeks before Trump's re-election noted that Biden's rhetorical support for Taiwan was not reflected in US arms sales data compared with his predecessors. According to the council's data, Trump and George W. Bush sold Taiwan the most arms in dollar terms since 1993:[10]

Clinton (1993–2001) – US$8.7 billion over 8 years

Bush (2001– 2009) – US$15.6 billion over 8 years

Obama (2009–2017) – US$13.9 billion over 8 years

Trump (2017– 2021) – US$18.3 billion over 4 years

Biden (2021–2025) – US$5.7 billion over 4 years

Before his re-election, Trump and some key appointees put forth their view that Taiwan needs to increase its defence spending dramatically. But at a time when countries are looking to see how serious Taiwan is about its defence, the KMT-led majority coalition in the Legislative Yuan made unprecedentedly large cuts to the national budget submitted by the Lai administration.

In the past, when lawmakers trimmed some of a proposed budget before passing it, the cuts would be around 1 per cent of the budget. The budget passed by the KMT and TPP, however, cut more than US$6.2 billion, or 6.6 per cent, from its total. (At the time of writing, the budget had been rejected by President Lai, who can only do so once.) The cuts were made across government agencies, provoking widespread public backlash. Even the military and diplomacy budgets were reduced at a time when both are crucial to Taipei's ability to rally its friends in Washington, Tokyo and elsewhere. Funds for military equipment and facilities were cut by 3 per cent, while half of the foreign ministry's budget was frozen.[11] As of March 2025, the budget was still being negotiated.

The KMT's rationale that the cuts were necessary due to government wastefulness was somewhat undercut by the fact that the proposed budget could be paid for with record projected government tax revenues. It was austerity, imposed during a time of surplus – and growing danger from across the strait. In mid-February 2025, five days before Trump's inauguration, Lai announced that he would draw up a special defence budget that would boost spending above 3 per cent of Taiwan's GDP.[12] Surprising no one, the KMT criticized the announcement.

Many DPP politicians see the hand of Beijing in the KMT-led coalition's moves since taking control of the legislature in early 2024. Among them is legislator Wang Ting-yu, Chair of the Foreign Affairs and National Defense Committee: 'The Chinese Communist Party is using its vast resources to secure control of key persons in the KMT,' Wang told me. 'Through those people they can exert control over the Legislative Yuan and inflict damage upon Taiwan's democracy.'

Among the weapons in the CCP's arsenal to influence the KMT, he said, were 'business favours, meetings with top Chinese officials, and kompromat'. The result is that now 'China can guide legislators to pass or eliminate laws in a way that helps China'.

At the top of those working for China's goals, Wang said, was the leader of the KMT legislative caucus, Fu Kun-chi. The legislator from Hualien, who had previously been imprisoned for insider trading, started off his new role by leading a delegation of seventeen KMT lawmakers to Beijing for a meeting with Wang Huning, the architect of Xi's policy towards Taiwan. 'Fu Kun-chi is one hundred per cent an advocate for the CCP,' Wang told me without hesitation.

In January 2025, the unofficial Indian embassy in Taipei celebrated India's Republic Day with a reception at the Grand Hyatt Taipei. Diplomats from most of the major countries were present, including Russia, whose representative was conspicuously avoided by the hundreds of other attendees. Conversations with both Taiwanese and foreigners tended to drift towards the unprecedented challenges Taiwan is facing. While the future was still unwritten, there was no denying that Taiwan is in a very precarious position.

This is not the first time the people of Taiwan have found themselves in a corner, and it comes at a time when Taiwan is more visible than ever and vital to the global economy. The Lai administration's diplomatic strategy sees integrating its diplomacy with its economic and technological prowess as a way to bring friendly countries closer. Items on its agenda include developing a 'China-free' drone supply chain with democratic friends, or obtaining licences to produce weapons that a strained US military industrial complex is incapable of producing at present.

In January 2025, the first cohort of Taiwanese military conscripts to complete their service after Tsai Ing-wen raised the mandatory conscription time from four months to twelve were discharged.[13] (Her predecessor, Ma Ying-jeou, had reduced conscription from twelve months to four during his presidency.) The conscripts' training had focused on defending against beach invasions. Soldiers received more live-fire experience than previous cohorts, while also getting experience in operating drones and man-portable Stinger missiles.

That same month, undersea cables linking mainland Taiwan with its outlying island county of Matsu were severed. Internet was partially restored to the islands via backup communications, and the cable connection was restored within two days. The incident highlights Taiwan's ability to quickly learn from and adapt to weaknesses. A similar incident two years earlier, in which two cables were cut by Chinese vessels, resulted in a total internet blackout for 13,000 Matsu residents that lasted fifty days.

Small but significant progress such as better-trained conscripts and improved emergency response capabilities are important, but it is generally accepted that Taiwan would be incapable of fending off China on its own. Building upon existing relationships with the US and other friendly countries while also establishing new ties with others will be vital to discouraging adventurism by Xi. But time is of the essence, and this problem does not threaten only Taiwan, but the world at large.

Whether they like it or not, democracies around the world now have to choose what kind of Taiwan they want to deal with: a valued and reliable partner in maintaining global stability, or a blade held against their necks by the most powerful authoritarian state in human history. At this point in the game, choosing not to oppose China's goal of swallowing Asia's freest country is effectively tacit approval for Beijing's expansionist agenda.

It's an unpleasant situation, and one that is being forced upon us all by a China bent on imposing its alternate reality regarding Taiwan on the rest of the world. Taiwan and China were once on opposite sides of a bitter Cold War rivalry, but Taiwan is no longer threatening its larger neighbour. It simply wants to keep the democracy that it fought for over generations.

It is important to recognize that the tensions surrounding Taiwan all spring from Beijing's illegitimate claim to be its government. As the open letter to Xi by Taiwan's Indigenous peoples noted in 2019, the Taiwanese are open to a peaceful, friendly coexistence, but only if Beijing is:

> *If one day China abandons its distorted understanding of history, nationality, and statehood;*
> *If one day China becomes our friendly neighbour and stops claiming its forceful 'parenthood' upon us;*
> Only then will we propose a toast to China, our neighbour, with a cup of millet wine in all our sincerity.[14]

ACKNOWLEDGEMENTS

There are countless people who helped make *Ghost Nation* possible – too many to thank here, but I will do my best.

I have been the grateful recipient of advice, assistance, insights, inspiration, food, drinks and other forms of support from innumerable people during my decade in Taiwan.

I would like to first thank my parents James and Valerie, my brother Ryan, my sister-in-law Kyla and my nephew Maddox. I am also grateful for the support of my many aunts, uncles and cousins.

I also received invaluable guidance and encouragement from my agent Marysia Juszczakiewicz at Peony Literary Agency, and my editors at Pan Macmillan: Mike Harpley and Ríbh Brownlee.

There is no way to acknowledge all the other people, organizations and establishments who have contributed directly or indirectly to the creation of this book. Below is a list that is far from exhaustive – please forgive me for any omissions. Given the state of affairs in China, I do not include my friends and colleagues in China and Hong Kong, but you should know who you are.

With that, I offer my heartfelt thanks to: Ami Bera, Another Brick, Nick Aspinwall, David Bandurski, Eveline Bingaman, Ben Blanchard, Bobwundaye, Patrick Boehler, Gerald Brown, Chris Buckley, Kathleen Calderwood, Argin Chang, Dennis Chang, Eric Chang, Wayne Chang, Amy Chang Chien, Vincent Chao, Steven Chase, Thompson Chau, Alicia Chen, Chen Hui-ling, Karissa

Chen, Kimberley Chen, Yu-Jie Chen, Chen-Yu Lin, Kathy Cheng, Han Cheung, Vic Chiang, Annabelle Chih, Josh Chin, Alysa Chiu, Kassy Cho, Catherine Chou, Claire Chu, Stephen Chua, Jerome Cohen, Elise Coker, Jeff Connelly, Cyrus Console, Paula Console-Soican, Fraser Crichton, Tim Culpan, Dan Curry, Chason Dailey, Helen Davidson, Becky Davis, David Demes, Chao Deng, Mike Denison, Lauren Dickey, James Donald, Gerry Doyle, Jessica Drun, Eight Immortals Grill, Samson Ellis, EPHK, Martin Fackler, Michael Fahey, Laura Fan, Emily Feng, Feng Shengli, Zsuzsa Anna Ferenczy, Tim Ferry, Mike Forsythe, Dave Frazier, Kyla Friend Horton, Nick Frisch, Michael Garber, Dinah Gardner, Brett Gerson, Bonnie Glaser, Will Glasgow, Jeremy Goldkorn, Nick Haggerty, Erin Hale, Fabian Hamacher, Rupert Hammond-Chambers, Mark Harrison, Kathrin Hille, Brian Hioe, Ryan Ho Kilpatrick, Vanessa Hope, Feliciana Hsu, Iris Hsu, Jason Hsu, Ilham Issak, Jakub Janda, Florian Janssen, Ward Johnsmeyer, Wu'er Kaixi, Ivan Kanapathy, Daniel Kao, Dean Karalekas, Adrian Kennedy, Yin Khvat, Tom Kirkpatrick, Kolas Yotaka, Katrina Ku, Lily Kuo, Michelle Kuo, Rhoda Kwan, Sebastien Lai, Ai-Men Lau, Peter Lavelle, Alex Lee, Daphne Lee, Jasmine Lee, Lee Jian Ting, Lee Lieh, Lee Teng-hui, Lee Tun-hou, Yian Lee, Ben Lewis, Maggie Lewis, Li Imte, Lisa Liang, Wen Lii, Amber Lin, Chen-Yu Lin, Lin Fei-fan, James Lin, Lin Miao-jung, Mindy Lin, John Liu, Juliet Lu, Myra Lu, Ted Lynch, Alec Martin, James Mayger, Nate Maynard, Simina Mistreanu, David Moll, Moonlab Foods, Paul Mozur, Gerry Mullany, Rich Murray-Nobbs, Steve Myers, Lev Nachman, Andrew Nagorski, Alexa Oleson, Jojje Oleson, Jeremy Olivier, Joanne Ou, Phil Pan, Alan Parkhouse, Hugo Peng, Matt Pottinger, Amy Qin, Austin Ramzy, Prashant Rao, Red Room Rendezvous, Revolver, Jane Rickards, Shelley Rigger, Tim Rinaldi, Lloyd Roberts, Gwen Robinson, Nadège Rolland, Alfred Romann, Roxy 36, Don Shapiro, Andy Sharp, Christian Shepherd, Eric Shih, Gerry Shih, Shrine Bar, Matthew Sills, Nicci Smith, Oğuz Solak, Jason Solana, Isabella Steger, Alice Su, Su Beng,

Acknowledgements

Chiaoning Su, Su Yen, Taiwan Foreign Correspondents' Club, Jane Tang, Chris Taylor, Kharis Templeman, Cat Thomas, Mark Thomas, Manohar Thyagaraj, Toasteria, Sarah Topol, Tsai Ing-wen, Lillian Tsay, Michael Turton, Sander Van de Moortel, Rosaline Walters, Ann Wang, Cindy Wang, Jenny Wang, Joyu Wang, Rosa Watmough, Clarissa Wei, Grace Weng, Rupert Wingfield-Hayes, Ed Wong, Florian Wong, Gillian Wong, Albert Wu, Emily Wu, Emily Pearson Wu, Enoch Wu, Joseph Wu, Sarah Wu, Wyatt's, An Rong Xu, Jenn Yang, Maysing Yang, Stephanie Yang, Will Yang, William Yang, Miracle Yeh, Michelle Yun, Sara Yurich and Matt Zaklad.

Thank you all. I dedicate this book to you.

LIST OF ACRONYMS

ADIZ   Air Defence Identification Zone
AI   Artificial Intelligence
AIT   American Institute in Taiwan
APEC   The Asia-Pacific Economic Cooperation
ARATS   Association for Relations Across the Taiwan Straits
CCP   Chinese Communist Party
CCTV   China Central Television
CSSTA   Cross-Strait Service Trade Agreement
DPP   Democratic Progressive Party. One of Taiwan's two dominant political parties, founded in opposition to the Kuomintang in 1986.
EU   European Union
EUV   Extreme ultraviolet lithography
FAPA   Formosan Association for Public Affairs
GDP   Gross Domestic Product
INDSR   Institute for National Defense and Security Research
IPAC   Inter-Parliamentary Alliance on China
ISO   International Organization for Standardization
ITRI   Industrial Technology Research Institute
KMT   Kuomintang. The Chinese Nationalist Party, one of Taiwan's two dominant political parties, which formerly ruled Taiwan as an authoritarian one-party state.
NATO   North Atlantic Treaty Organization

PLA	People's Liberation Army
PRC	People's Republic of China
ROC	Republic of China
SEF	Straits Exchange Foundation
TPP	Taiwan's People Party
TRA	Taiwan Relations Act
TSMC	Taiwan Semiconductor Manufacturing Company
UMC	United Microelectronics Corporation
UN	United Nations
VOC	Dutch East India Company (Vereenigde Oostindische Compagnie in Dutch)
WHA	World Health Assembly
WHO	World Health Organization
WTO	World Trade Organization
WUFI	World United Formosans for Independence

NOTES

INTRODUCTION — DARK CLOUDS

1 Wang, Flor and Pan Tzu-yu, 'Taiwan on course to become super-aged society by 2025', *Focus Taiwan* (17 October 2024) [https://focustaiwan.tw/society/202410170030]
2 Wu Hsin-yun and Lee Hsin-Yin, 'Mid-level workers most needed amid Taiwan's labor shortage: Minister', *Focus Taiwan* (30 October 2024) [https://focustaiwan.tw/society/202410300018]; Lin, Sean, 'MND aims to stem exodus of volunteer soldiers with pay raises', *Focus Taiwan* (15 January 2025) [https://focustaiwan.tw/politics/202501150018]
3 Gan, Nectar, 'Who is Lai Ching-te, Taiwan's new President?', CNN (13 January 2024) [https://edition.cnn.com/2024/01/14/asia/profile-lai-ching-te-taiwan-new-president-intl-hnk/index.html]
4 Takahashi, Tetsushi, 'China deploys new missile seen as "Guam Express"', *Nikkei Asia* (18 April 2018) [https://asia.nikkei.com/Politics/China-deploys-new-missile-seen-as-Guam-Express]
5 Welch, Jennifer; Leonard, Jenny; Cousin, Maeva; DiPippo, Gerard; and Orlik, Tom, 'Xi, Biden and the $10 Trillion Cost of War Over Taiwan', Bloomberg News (9 January 2024) [https://www.bloomberg.com/news/features/2024-01-09/if-china-invades-taiwan-it-would-cost-world-economy-10-trillion]
6 Chau, Thompson, 'Taiwan foreign minister vows to work with Trump on "democratic supply chain"', *Nikkei Asia* (7 January 2025) [https://asia.nikkei.com/Editor-s-Picks/Interview/Taiwan-

foreign-minister-vows-to-work-with-Trump-on-democratic-supply-chain]

ONE — APPEASING THE AFTERWORLD

1 'President Tsai apologizes to indigenous peoples on behalf of government', Office of the President, Republic of China (Taiwan) (1 August 2016) [https://english.president.gov.tw/NEWS/4950]
2 '原住民族人口及健康統計年報', 原住民族委員會 [https://cip.nhri.edu.tw/annual_report/stat/pop]
3 Xi Jinping, 'Working Together to Realize Rejuvenation of the Chinese Nation and Advance China's Peaceful Reunification', Taiwan Work Office of the CPC Central Committee (12 April 2019) [http://www.gwytb.gov.cn/wyly/201904/t20190412_12155687.htm]
4 Aspinwall, Nick, 'Taiwan's Indigenous to Xi: Taiwan Is Not China, Xi Does Not "Understand Dignity"', *The News Lens* (9 January 2019) [https://international.thenewslens.com/article/111667]
5 '原轉會各民族代表：台灣原住民族致中國習近平主席' [https://gov.hackmd.io/@chihao/SyKTh6bM4?type=view]
6 Liu Jiun-Yu, 'Intertwined maritime Silk Road and Austronesian routes: A Taiwanese archaeological perspective', *Journal of Global History* (Cambridge University Press, 9 October 2023) [https://www.cambridge.org/core/journals/journal-of-global-history/article/intertwined-maritime-silk-road-and-austronesian-routes-a-taiwanese-archaeological-perspective/7517F668B827276F33E112DEEC944311]
7 Wu, Albert and Kuo, Michelle, 'Ghost Month Is Almost Here', *The News Lens* (14 August 2023) [https://international.thenewslens.com/article/186474]
8 Wu and Kuo, 'Ghost Month Is Almost Here'

TWO — ILHA FORMOSA AND ITS COLONIZERS

1 Taiwan Public Television, '賴清德總統就職國宴首度移師台南開近百桌' (21 May 2024) [https://www.google.com/url?sa=t&source=web&rct=j&opi=89978449&url=https://www.youtube.com/watch%3Fv%3DoDZhKH0Bx2k&ved=2ahUKEwjv7rOLkpqIAxWkna8BHbRGDmQQz4oFegQIDRAK&usg=AOvVaw07KjDZGXwsl4lcsmVW6SwL]

2   Wei, Clarissa, 'Taiwan's Dinner Table Diplomacy', *Foreign Policy* (19 May 2024) [https://foreignpolicy.com/2024/05/19/taiwan-lai-ching-te-president-inauguration-banquet-food-china-culture-democracy/]

3   Staff, '賴清德：國際社會公認台海緊張升高根源在中國', Radio France Internationale (25 August 2023)

4   Hernandez, Javier C., 'Latest Clash Between Taiwan and China: A Fistfight in Fiji', *The New York Times* (19 October 2020) [https://www.nytimes.com/2020/10/19/world/asia/china-taiwan-fiji-fight.html]

5   Wingfield-Hayes, Rupert, 'China's Rhetoric Turns Dangerously Real for Taiwanese', BBC News (16 August 2024) [https://www.bbc.com/news/articles/ce8dy437pdno]

6   José Eugenio Borao Mateo, *Spaniards in Taiwan: Documents,* 2 vols. (Taipei: SMC, 2001), 1:1–15

7   Van der Wees, Gerrit, 'How European Explorers Discovered Ilha Formosa', *Taipei Times* (13 February 2019) [https://www.taipeitimes.com/News/feat/archives/2019/02/13/2003709641]

8   Andrade, Tonio, *Lost Colony: The Untold Story of China's First Great Victory Over the West* (Princeton University Press, 2011), p. 26

9   Andrade, *Lost Colony*, p. 27

THREE — IN THE SHADOWS OF EMPIRE

1   Cheung, Han, 'Taiwan in Time: The Swede Who Lost Formosa', *Taipei Times* (28 January 2018) [https://www.taipeitimes.com/News/feat/archives/2018/01/28/2003686554]

2   Teng, Cathy, 'Island Storytellers: Trees in Contemporary Taiwan', *Taiwan Panorama* (May 2023) [https://www.taiwan-panorama.com/en/Articles/Details?Guid=6820f873-c180-4560-b981-d85f50de661b&CatId=10&postname=Island%20Storytellers%3A%20-Trees%20in%20Contemporary%20Taiwan]

3   Teng, Emma, 'Taiwan and Modern China', *Oxford Research Encyclopedias* (23 May 2019) [https://oxfordre.com/asianhistory/display/10.1093/acrefore/9780190277727.001.0001/acrefore-9780190277727-e-155]

4   Teng, 'Taiwan and Modern China'

5   Staff, 'The Pirates of Formosa', *The New York Times* (24 August 1867)

6 Cheung, Han, 'Mudan Incident: Pivotal Point in Asian History', *Taipei Times* (31 May 2022) [https://www.taipeitimes.com/News/feat/archives/2022/05/31/2003779057]

7 Staff, 'Savage Island of Formosa Transformed by Japanese', *The New York Times* (25 September 1904)

8 Cheung, Han, 'Taiwan in Time: Magic amulets, tax breaks and a messiah', *Taipei Times* (2 July 2017) [https://www.taipeitimes.com/News/feat/archives/2017/07/02/2003673711]

9 Kerr, George, *Formosa Betrayed* (Camphor Press, 2017, 3rd edition) [Kindle edition], chapter 1, loc. 595

FOUR — SHIFTING WINDS, DEADLY STORM

1 Cheung, Han, 'Taiwan in Time: Anticolonial Messages from the Sky', *Taipei Times* (16 October 2016) [https://www.taipeitimes.com/News/feat/archives/2016/10/16/2003657251]

2 Cheung, Han, 'Photo exhibition: Weaving Taiwan's story through photos', *Taipei Times* (10 August 2021) [https://www.taipeitimes.com/News/feat/archives/2021/08/10/2003762337]

3 Kerr, George, *Formosa Betrayed* (Camphor Press, 2017, 3rd edition) [Kindle edition], chapter 2, loc. 952

4 Kerr, *Formosa Betrayed*, chapter 2, loc. 964

5 Kerr, *Formosa Betrayed*, chapter 3, loc. 1494–519

6 'Press Communique Regarding Conference of President Roosevelt, Generalissimo Chiang Kai-shek, and Prime Minister Churchill at Cairo', U.S. Department of State Office of the Historian [https://history.state.gov/historicaldocuments/frus1943China/d136; https://digitalarchive.wilsoncenter.org/document/cairo-declaration]

7 Kerr, *Formosa Betrayed*, chapter 3, loc. 1454

8 Kerr, *Formosa Betrayed*, chapter 3, loc. 1480

9 Kerr, *Formosa Betrayed*, ibid.

10 Kerr, *Formosa Betrayed*, ibid.

11 Cheung, Han, 'Taiwan in Time: Taiwan's 'Great Leap Forward', *Taipei Times* (9 October 2016)

12 Kerr, *Formosa Betrayed*, chapter 5, loc. 1846

13 Cheung, Han, 'Taiwan in Time: The Iron Councilor's Demands', *Taipei Times* (25 February 2018)

14 Lin, Adela and Ellis, Samson, 'Once Worth Billions, Long-Ruling Taiwan Party Now Short of Cash', Bloomberg News (27 September 2017) [https://www.bloomberg.com/politics/articles/2017-09-27/once-worth-billions-long-ruling-taiwan-party-now-short-of-cash]
15 Kerr, *Formosa Betrayed*, chapter 5, loc. 1784
16 'Plague', Taiwan Centers for Disease Control [https://www.cdc.gov.tw/En/Category/ListContent/bgog_VU_Ysrgkes_KRUDgQ?uaid=iKDX_K5V86LsGGBkl5Em-Q]
17 戰役之後, Academia Sinica Archives of the Institute of Taiwan History [https://archives.ith.sinica.edu.tw/collections_con.php?no=287]
18 Kerr, *Formosa Betrayed*, chapter 9, loc. 3100
19 Kerr, *Formosa Betrayed*, chapter 9, loc. 3225
20 Horton, Chris, 'Taiwan Commemorates a Violent Nationalist Episode, 70 Years Later', *The New York Times* (26 February 2017)
21 Horton, 'Taiwan Commemorates a Violent Nationalist Episode'
22 Staff, 'These are the Tyrants and Robber Barons of the 228 Massacre', *The Reporter* (20 February 2017)
23 Kerr, *Formosa Betrayed*, chapter 14, loc. 4567
24 Durdin, Tillman, 'FORMOSA KILLINGS ARE PUT AT 10,000: Foreigners Say the Chinese Slaughtered Demonstrators Without Provocation', *The New York Times* (29 March 1947) [https://www.nytimes.com/1947/03/29/archives/formosa-killings-are-put-at-10000-foreigners-say-the-chinese.html]
25 Kerr, *Formosa Betrayed*, chapter 14, loc. 4648
26 Horton, 'Taiwan Commemorates a Violent Nationalist Episode'
27 Staff, 'Chiang bears ultimate responsibility for 228 Incident: scholars', *Focus Taiwan* (23 February 2017) [https://focustaiwan.tw/culture/201702230028]
28 Chen, Patrick and Butler Stas, 'Protestors Interrupt Taipei Mayor's 228 Incident Speech', TaiwanPlus News (28 February 2023) [https://www.youtube.com/watch?v=CLCoDboLqYk]

FIVE — WHITE TERROR IN FREE CHINA

1 Kerr, George, *Formosa Betrayed* (Camphor Press, 2017, 3rd edition), chapter 18, loc. 5895

2. Hsieh, Wen-hua, 'Letters from the White Terror Era', *Taipei Times* (15 July 2009) [https://www.taipeitimes.com/News/taiwan/archives/2009/07/15/2003448722]
3. Mozur, Paul, 'Taiwan Families Receive Goodbye Letters Decades After Executions', *The New York Times* (3 February 2016) [https://www.nytimes.com/2016/02/04/world/asia/taiwan-white-terror-executions.html]
4. Interactive feature, 'Letters from the White Terror', *The New York Times* (3 February 2016) [https://www.nytimes.com/interactive/2016/02/03/world/asia/04taiwan-white-terror-execution-letters.html]
5. Interactive feature, 'Letters from the White Terror'
6. 謝聰敏, 談景美軍法看守所 (李敖出版社,1991), pp. 1–269 [https://books.google.com.tw/books/about/%E8%AB%87%E6%99%AF%E7%BE%8E%E8%BB%8D%E6%B3%95%E7%9C%8B%E5%AE%88%E6%89%80.html?id=lYJInQEACAAJ&redir_esc=y]
7. Yang, Pi-chuan, *The Road to Freedom: Taiwan's Postwar Human Rights Movement* (Taipei, Dr. Chen Wen-chen Memorial Foundation, 2004), p. 27
8. Cheung, Han, 'A Long Journey Full of Tears', *Taipei Times* (27 June 2013) [https://www.taipeitimes.com/News/feat/archives/2017/06/27/2003673373]
9. Cheung, 'A Long Journey Full of Tears'
10. Staff, 'FORMOSA: Man of the Single Truth', *Time* (18 April 1955) [https://time.com/archive/6803801/formosa-man-of-the-single-truth/]
11. Yang, *The Road to Freedom*, pp. 26–7
12. Horton, Chris, 'A Taiwan Museum Featuring All of Asia', *The New York Times* (6 October 2016) [https://www.nytimes.com/2016/10/06/arts/international/a-taiwan-museum-featuring-all-of-asia.html]
13. Dillery, Edward, 'Harvey Feldman', *The Association for Diplomatic Studies and Training, Foreign Affairs Oral History Project* (2001), p. 46 [https://www.adst.org/OH%20TOCs/Feldman,%20Harvey.toc.pdf]
14. Staff, 'Taiwan Said to Execute Man Who Wrote Peking', *The New York Times* (15 January 1971) [https://www.nytimes.com/1971/01/15/archives/taiwan-said-to-execute-man-who-wrote-peking.html]
15. Cheung, Han, 'Taiwan in Time: Burning for Freedom', *Taipei Times* (16 May 2021) [https://www.taipeitimes.com/News/feat/archives/2021/05/16/2003757476]

16   Staff, '35 years later, freedom of speech defender Nylon Cheng's legacy lives on', *Focus Taiwan* (7 April 2024) [https://focustaiwan.tw/politics/202404070008]

SIX — DEATH OF A DYNASTY

1   Staff, 'Taiwan: A Shot at Chiang', *Time* (4 May 1970) [https://time.com/archive/6876891/taiwan-a-shot-at-chiang/]
2   Lelyveld, Joseph, 'Entrance to the Plaza Hotel Is the Scene of Assassination Attempt', *The New York Times* (25 April 1970) [https://www.nytimes.com/1970/04/25/archives/entrance-to-the-plaza-hotel-is-the-scene-of-assassination-attempt.html]
3   Eckholm, Erik, 'Taipei Journal; Human Rights Stalwart Has an Unlikely Resume', *The New York Times* (13 June 2000) [https://www.nytimes.com/2000/06/13/world/taipei-journal-human-rights-stalwart-has-an-unlikely-resume.html]
4   Staff, 'Chiang Kai-shek is Dead in Taipei at 87; Last of Allied Big Four of World War II', *The New York Times* (6 April 1975) [https://www.nytimes.com/1975/04/06/archives/chiang-kaishek-is-dead-in-taipei-at-87-last-of-allied-big-four-of.html]
5   Cohen, Jerome A., *Eastward, Westward* (Columbia University Press, 2025) [Kindle edition], chapter 20, loc. 4729
6   'The Murder of Henry Liu: Hearings and Markup Before the Committee on Foreign Affairs and its Subcommittee on Asian and Pacific Affairs', House of Representatives Ninety Ninth Congress First Session (1985), p. 48
7   'The Murder of Henry Liu: Hearings and Markup', p. 48
8   'The Murder of Henry Liu: Hearings and Markup', p. 47
9   'The Murder of Henry Liu: Hearings and Markup', p. 2
10   Adam, John, 'Taiwanese Here Fear Murder', *The Michigan Daily* (9 July 1981), p. 1
11   Peterson, Bill, 'After Police Interrogation, A Death', *The Washington Post* (27 July 1981) [https://www.washingtonpost.com/archive/politics/1981/07/28/after-police-interrogation-a-death/24a5b464-c563-4546-a3da-353b34c0526a/]

12 Kamm, Henry, 'Taipei Irritated by Protests in US', *The New York Times* (9 August 1981) [https://www.nytimes.com/1981/08/09/world/taipei-irritated-by-protests-in-us.html]
13 Wecht, Cyril H., 'Murder in Taiwan', *The American Journal of Forensic Medicine and Pathology* (June 1985), pp. 97–104
14 'The Murder of Henry Liu: Hearings and Markup', pp. 4–5
15 'The Murder of Henry Liu: Hearings and Markup', p. 10
16 'The Murder of Henry Liu: Hearings and Markup', p. 36
17 Cohen, *Eastward, Westward*, chapter 20, loc. 4767
18 Cohen, *Eastward, Westward*, chapter 20, loc. 4854
19 Hille, Kathrin, 'Killer's death haunts Taiwan party', *Financial Times* (3 November 2007) [https://www.ft.com/content/0941d960-8973-11dc-b52e-0000779fd2ac]

SEVEN — THE THREE IMMORTALS

1 Pace, Eric, 'Chiang Ching-kuo Dies at 77, Ending a Dynasty on Taiwan', *The New York Times* (14 January 1988) [https://www.nytimes.com/1988/01/14/obituaries/chiang-ching-kuo-dies-at-77-ending-a-dynasty-on-taiwan.html]
2 Cheung, Han, 'Taiwan in Time: Awakening a silenced past', *Taipei Times* (26 February 2017) [https://www.taipeitimes.com/News/feat/archives/2017/02/26/2003665718]
3 Hickey, Dennis V., *Foreign Policy Making in Taiwan* (Routledge, 2006), p. 88
4 Chau, Thompson, '"We are Chinese": KMT lawmaker challenges president on Taiwan's nature', *Nikkei Asia* (1 October 2024) [https://t.co/Jruz9cU2yb]
5 '1996 總統大選彭明敏第一次電視政見會', 彭明敏文教基金會 (First Televised Political Platform Presentation of the 1996 Presidential Election, Peng Ming-min Cultural and Education Foundation) [https://www.youtube.com/watch?v=-icSnQP9U1I]
6 Peng Ming-Min, *A Taste of Freedom: Memoirs of a Taiwanese Independence Leader* (Camphor Press, 2017) [Kindle edition], chapter 2, loc. 601
7 Peng Ming-Min, *A Taste of Freedom*, chapter 2, loc. 682–92
8 Peng Ming-Min, *A Taste of Freedom*, chapter 8

9. Horton, Chris, 'Peng Ming-min, Fighter for Democracy in Taiwan, Dies at 98', *The New York Times* (16 April 2022) [https://www.nytimes.com/2022/04/16/world/asia/peng-ming-min-dead.html]
10. Staff, 'Taiwan recognizes Dutch lawyer with prestigious award', *Taipei Times* (2 October 2023) [https://www.taipeitimes.com/News/taiwan/archives/2023/10/02/2003807090]

EIGHT – THE DEMOCRACY EXPERIMENT

1. Lin, Irene, 'Kaohsiung Eight trial pointed way to Taiwan's future', *Taipei Times* (9 December 1999) [https://www.taipeitimes.com/News/local/archives/1999/12/09/0000014182]
2. Lin, 'Kaohsiung Eight trial pointed way to Taiwan's future'
3. Platt, Kevin, 'A Chinese Leader Aims to Be One of the Folks', *The Christian Science Monitor* (21 July 1997) [https://www.csmonitor.com/1997/0721/072197.intl.intl.1.html]
4. Shih Hsiiu-chuan, 'Su Chi admits the "1992 consensus" was made up', *Taipei Times* (22 February 2006) [https://www.taipeitimes.com/News/taiwan/archives/2006/02/22/2003294106]
5. Hsieh Yi-hsuan; Lu Chia-jung; Li Ya-wen; and Lai, Sunny, 'China urged to be "pragmatic" after rejecting new cross-strait framework idea', *Focus Taiwan* (16 October 2024) [https://focustaiwan.tw/cross-strait/202410160023]
6. Staff, 'President details missile threat', *Taipei Times* (2 December 2003) [https://www.taipeitimes.com/News/taiwan/archives/2003/12/02/2003078033]
7. Knowlton, Brian, 'Bush warns Taiwan to keep status quo: China welcomes U.S. stance', *International Herald Tribune* (10 December 2003) [https://www.nytimes.com/2003/12/10/news/bush-warns-taiwan-to-keep-status-quo-china-welcomes-us-stance.html]
8. Chang Yun-ping, 'Two million rally for peace', *Taipei Times* (28 February 2004) [https://www.taipeitimes.com/News/front/archives/2004/02/29/2003100533]
9. Watts, Jonathan, 'Taiwan president shot in election attack', *The Guardian* (20 March 2004) [https://www.theguardian.com/world/2004/mar/20/china.jonathanwatts]

10 Buckley, Chris and Chang Chien, Amy, 'Shih Ming-teh, Defiant Activist for a Democratic Taiwan, Dies at 83', *The New York Times* (23 January 2024) [https://www.nytimes.com/2024/01/23/world/asia/shih-ming-teh-dead.html]

11 Buckley and Chang Chien, 'Shih Ming-teh, Defiant Activist for a Democratic Taiwan, Dies at 83'

12 Bradsher, Keith, 'Protestors Fuel a Long-Shot Bid to Oust Taiwan's Leader', *The New York Times* (28 September 2006) [https://www.nytimes.com/2006/09/28/world/asia/28taiwan.html]

13 Kagan, Richard C., and Haggerty, Nicholas, 'Time to grant Chen Shui-bian a pardon', *Taipei Times* (12 April 2024) [https://www.taipeitimes.com/News/editorials/archives/2024/04/12/2003816293]

14 Chen Feng-li, 'Funeral to Honor Self-immolator', *Taipei Times* (3 January 2009) [https://www.taipeitimes.com/News/taiwan/archives/2009/01/03/2003432764]

15 Mo Yan-chih, 'Cross-strait service trade pact signed', *Taipei Times* (22 Jun 2013) [https://www.taipeitimes.com/News/front/archives/2013/06/22/2003565371]

16 Davidson, Helen, 'How the Sunflower movement birthed a generation determined to protect Taiwan', *The Guardian* (21 March 2024) [https://www.theguardian.com/world/2024/mar/21/what-is-taiwan-sunflower-movement-china]

17 Hioe, Brian, 'Sunflower Movement Activists Found Not Guilty', *New Bloom* (31 March 2017) [https://newbloommag.net/2017/03/31/sunflower-movement-not-guilty/]

NINE — TAIWAN IN THE BALANCE

1 Gan, Nectar, 'Xi vows "reunification" with Taiwan on eve of Communist China's 75th birthday', CNN (1 October 2024) [https://edition.cnn.com/2024/10/01/china/china-xi-reunification-taiwan-national-day-intl-hnk/index.html]

2 Hitler, Adolf, *Mein Kampf (My Struggle)* (Hurst and Blackett Ltd., 1939), p. 1 [https://gutenberg.net.au/ebooks02/0200601h.html]

3 Blanchard, Ben, '"Impossible" for People's Republic of China to be our motherland, Taiwan president says', Reuters (5 October 2024)

[https://www.reuters.com/world/asia-pacific/impossible-peoples-republic-china-be-our-motherland-taiwan-president-says-2024-10-05/]

4 Blanchard, Ben, 'If China wants Taiwan it should also take back land from Russia, president says', Reuters (2 September 2024) [https://www.reuters.com/world/asia-pacific/if-china-wants-taiwan-it-should-also-take-back-land-russia-president-says-2024-09-02/]

5 Lai Yu-chen; Su Ssu-yun; Tseng Jen-kai; and Huang, Frances, 'Taiwan's government condemns cyberattacks on TWSE, DGBAS, financial firms', Focus Taiwan (13 September 2024) [https://focustaiwan.tw/politics/202409130016]

6 'President Lai delivers 2024 National Day Address', Office of the President, Republic of China (Taiwan) (10 October 2024) [https://english.president.gov.tw/NEWS/6816]

7 'Public Opinion Back's President's Cross-Strait Policy Stances in National Day Address, Rejects CCP Political Positions and Unrelenting Pressure on Taiwan', Mainland Affairs Council, Republic of China (Taiwan) (24 October 2024) [https://www.mac.gov.tw/en/News_Content.aspx?n=A921DFB2651FF92F&sms=37838322A6DA5E79&s=0B98923AF32A235E]

8 Hille, Kathrin, 'China's show of force in massive military exercises alarms Taiwan', Financial Times (17 October 2024)

9 Funaiole, Matthew P. and Bermudez Jr., Joseph S., 'China's New Amphibious Assault Ship Sails into the South China Sea', Center for Strategic & International Studies (24 November 2020) [https://www.csis.org/analysis/chinas-new-amphibious-assault-ship-sails-south-china-sea]

10 Funaiole, Matthew P.; Hart, Brian; Powers-Riggs, Aidan; and Bermudez Jr., Joseph S, 'China's Massive Next-Generation Amphibious Assault Ship Takes Place', Center for Strategic & International Studies (1 August 2024) [https://www.csis.org/analysis/chinas-massive-next-generation-amphibious-assault-ship-takes-shape]

11 Horton, Chris, 'China mobilizes civilian ferries for Taiwan invasion drills', Nikkei Asia (25 August 2021) [https://asia.nikkei.com/Politics/International-relations/China-mobilizes-civilian-ferries-for-Taiwan-invasion-drills]

12 Zakaria, Fareed, 'On GPS: Is Taiwan's tech under threat?', CNN (31 July 2022) [https://edition.cnn.com/videos/tv/2022/07/31/exp-731-taiwan-tech-mark-liu-tsmc.cnn]

13. '中國文攻新招？ 洗腦歌「坐上動車去台灣」 誇稱2035將統一', Liberty Times Net (6 November 2021) [https://news.ltn.com.tw/news/politics/breakingnews/3728461]
14. Sullivan, Rory, 'Beijing will "re-educate" Taiwan if it takes over island, Chinese ambassador warns', *The Independent* (4 August 2022) [https://www.independent.co.uk/independentpremium/world/nancy-pelosi-china-taiwan-invasion-b2138314.html]
15. Tillett, Andrew, 'China plans re-education "once Taiwan is united"', *Australia Financial Review* (10 August 2022) [https://www.afr.com/politics/federal/no-room-for-compromise-over-taiwan-china-envoy-20220810-p5b8pz]
16. 'Xinjiang: China defends "education" camps', BBC News (17 September 2020) [https://www.bbc.com/news/world-asia-china-54195325]
17. 'One-minute handshake marks historic meeting between Xi Jinping and Ma Ying-jeou', *The Straits Times* (19 January 2016) [https://www.straitstimes.com/singapore/one-minute-handshake-marks-historic-meeting-between-xi-jinping-and-ma-ying-jeou]
18. 'Opening remarks by President Ma Ying-jeou at his meeting with mainland Chinese leader Xi Jinping in Singapore', Office of the President, Republic of China (Taiwan) (7 November 2015) [https://english.president.gov.tw/NEWS/4779]
19. Perlez, Jane, 'In Heat of August 1945, Mao and Chiang Met for the Last Time', *The New York Times* (4 November 2015) [https://www.nytimes.com/2015/11/05/world/asia/the-last-time-mao-and-chiang-met.html]
20. Horton, Chris, 'The Taiwanese Populist Advancing China's Interests', *The Atlantic* (16 April 2019) [https://www.theatlantic.com/international/archive/2019/04/taiwanese-populist-han-kuo-yu-china/587146/]
21. Horton, 'The Taiwanese Populist Advancing China's Interests'
22. '韓國瑜會晤國台辦主任劉結一 雙方釋出善意!,' CTI (25 March 2019) [https://www.youtube.com/watch?v=8lZjapExnvI]
23. '韓國瑜會晤國台辦主任劉結一 雙方釋出善意!,' CTI
24. 'Taiwanese/Chinese Identity (1992/06-2024/12)', Election Study Center, National Chengchi University (13 January 2025) [https://esc.nccu.edu.tw/PageDoc/Detail?fid=7800&id=6961]

25 Pan, Jason, 'Tsai's approval rating rising, poll shows', *Taipei Times* (20 May 2019) [https://www.taipeitimes.com/News/taiwan/archives/2019/05/20/2003715462]

26 'Xi Jinping says Taiwan "must and will be' reunited with China", BBC News (3 January 2019) [https://www.bbc.com/news/world-asia-china-46733174]

27 Horton, Chris, 'Taiwan's President, Defying Xi Jinping, Calls Unification Offer "Impossible"', *The New York Times* (5 January 2019) [https://www.nytimes.com/2019/01/05/world/asia/taiwan-xi-jinping-tsai-ing-wen.html]

28 Chau, Thompson, '"We are Chinese": KMT lawmaker challenges president on Taiwan's nature', *Nikkei Asia* (1 October 2024) [https://asia.nikkei.com/Editor-s-Picks/Interview/We-are-Chinese-KMT-lawmaker-challenges-president-on-Taiwan-s-nature]

TEN — SCRUBBING TAIWAN'S SOVEREIGNTY

1 UN General Assembly, 'Restoration of the lawful rights of the People's Republic of China in the United Nations', United Nations Digital Library (26th Session, 25 October 1971) [https://digitallibrary.un.org/record/192054?ln=en&v=pdf]

2 'Daily Press Briefing by the Office of the Spokesperson for the Secretary-General', United Nations (27 March 2023) [https://press.un.org/en/2023/db230327.doc.htm]

3 Horton, Chris, 'Is China Running the UN?', *The China Project* (5 September 2023) [https://thechinaproject.com/2023/09/05/is-china-running-the-un/]

4 Tanner, Henry, 'U.N. Seats Peking and Expels Taipei', *The New York Times* (26 October 1971) [https://archive.nytimes.com/www.nytimes.com/library/world/asia/102671china-relations.html]

5 'IPAC Taipei Summit 2024 – Summit Report', Inter-Parliamentary Alliance on China (1 August 2024) [https://www.ipac.global/campaigns/ipac-taipei-summit-2024---summit-report]

6 'Australian Senate Unanimously Passes Urgency Motion on UN Resolution 2758', Inter-Parliamentary Alliance on China (21 August 2024) [https://www.ipac.global/campaigns/australian-senate-unanimously-passes-urgency-motion-on-un-resolution-2758]

7   Fang Wei-li; Chen Cheng-yu; and Yeh, Esme, 'MOFA thanks Dutch House for support', *Taipei Times* (14 September 2024) [https://www.taipeitimes.com/News/taiwan/archives/2024/09/14/2003823769]
8   'H.R. 1176 – Taiwan International Solidarity Act', Congress.gov (13 May 2023) [https://www.congress.gov/bill/118th-congress/house-bill/1176/text]
9   Lin Che-yuan and Hiciano, Lery, 'Talks on UN resolution statement break down', *Taipei Times* (9 October 2024) [https://www.taipeitimes.com/News/taiwan/archives/2024/10/09/2003825006]
10 Drun, Jessica and Glaser, Bonnie, 'The Distortion of UN Resolution 2758 and Limits on Taiwan's Access to the United Nations', The German Marshall Fund (24 March 2022) [https://www.gmfus.org/news/distortion-un-resolution-2758-and-limits-taiwans-access-united-nations]
11 'Joint Press Release – Support for Taiwan's Meaningful Engagement with the WHO and Participation as an Observer in the WHA', American Institute in Taiwan (24 May 2024) [https://www.ait.org.tw/joint-press-release-support-for-taiwans-meaningful-engagement-with-the-world-health-organizationand-participation-as-an-observer-in-the-wha/]
12 Quinn, Jimmy, 'Deputy U.N. Head Makes Rare Comments Backing Taiwan', *National Review* (15 September 2023) [https://www.nationalreview.com/corner/deputy-u-n-head-makes-rare-comments-backing-taiwan/]
13 Quinn, 'Deputy U.N. Head Makes Rare Comments Backing Taiwan'
14 Blanchard, Ben, 'US decries Nauru's "unfortunate" ditching of Taiwan, warns on China's promises', Reuters (16 January 2024) [https://www.reuters.com/world/asia-pacific/naurus-decision-break-ties-with-taiwan-unfortunate-us-official-2024-01-16/]
15 Myers, Steven Lee and Horton, Chris, 'China Tries to Erase Taiwan, One Ally (and Website) at a Time', *The New York Times* (25 May 2018) [https://www.nytimes.com/2018/05/25/world/asia/china-taiwan-identity-xi-jinping.html]
16 Information obtained on Taiwan and China via database search on visaindex.com in March 2025 [https://visaindex.com/visa-requirement/]
17 Oberdorfer, Don, 'Taiwanese Mob Pelts U.S. Diplomatic Team', *The Washington Post* (27 December 1978) [https://www.washingtonpost.com/

archive/politics/1978/12/28/taiwanese-mob-pelts-us-diplomatic-team/db87b934-225c-4794-a437-90edcc0dcbcc/]

18 Kamm, Henry, 'Taiwanese Attack U.S. Motorcade as Officials Arrive for Negotiations', *The New York Times* (28 December 1978) [https://www.nytimes.com/1978/12/28/archives/taiwanese-attack-us-motorcade-as-officials-arrive-for-negotiations.html]

19 Shapiro, Don, 'The Taiwan Relations Act, the Linchpin of Taiwan's Relations with the U.S.', *Taiwan Business Topics* (17 April 2019) [https://topics.amcham.com.tw/2019/04/taiwan-relations-act-the-linchpin-of-us-taiwan-relations/]

20 Kamm, 'Taiwanese Attack U.S. Motorcade' (28 December 1978)

21 Shapiro, 'The Taiwan Relations Act' (17 April 2019)

22 Blinken, Antony, 'On Taiwan's Election', U.S. Department of State (13 January 2024) [https://www.state.gov/on-taiwans-election/]

23 'Joint Statement Following Discussions with Leaders of the People's Republic of China', U.S. Department of State Office of the Historian (28 February 1972) [https://history.state.gov/historicaldocuments/frus1969-76v17/d203]

24 Lord, Winston, 'Conversations with Chou En-lai: July 10, afternoon sessions', The White House (6 August 1971) [https://nsarchive2.gwu.edu/NSAEBB/NSAEBB66/ch-35.pdf]

25 'Memorandum of Conversation', The White House (11 July 1971) [https://nsarchive2.gwu.edu/NSAEBB/NSAEBB66/ch-38.pdf]

26 Skidmore, David, 'What it Means to Be an "Old Friend of the Chinese People"', *The Diplomat* (12 December 2023) [https://thediplomat.com/2023/12/what-it-means-to-be-an-old-friend-of-the-chinese-people/]

27 Dillery, Edward, 'Harvey Feldman', The Association for Diplomatic Studies and Training (2001) [https://www.adst.org/OH%20TOCs/Feldman,%20Harvey.toc.pdf]

28 Bosco, Joseph, 'Reassessing the Mutual Defense Treaty: Three Communiques, Three Erosions of Taiwan's Interests', Global Taiwan Institute (10 January 2018) [https://globaltaiwan.org/2018/01/reassessing-the-mutual-defense-treaty-three-communiques-three-erosions-of-taiwans-interests/]

29 Taubman, Philip, 'U.S. and Peking Join in Tracking Missiles in Soviet', *The New York Times* (18 June 1981) [https://www.nytimes.

   com/1981/06/18/world/us-and-peking-join-in-tracking-missiles-in-soviet.html]
30 Lardner Jr., George and Smith, R. Jeffrey, 'Intelligence Ties Endure Despite U.S.-China Strain', *The Washington Post* (24 June 1989) [https://www.washingtonpost.com/archive/politics/1989/06/25/intelligence-ties-endure-despite-us-china-strain/f8b2789d-0f0c-4ea7-932b-9f4267a994a3/?_pml=1]
31 'U.S.-PRC Joint Communique (1982)', American Institute in Taiwan (17 August 1982)[https://www.ait.org.tw/u-s-prc-joint-communique-1982/]
32 'U.S.-PRC Joint Communique (1982)', American Institute in Taiwan
33 Reagan, Ronald, 'Arms Sales to Taiwan', The White House (17 August 1982)[https://china.usc.edu/ronald-reagan-arms-sales-taiwan-august-17-1982]
34 Shultz, George, 'Assurances for Taiwan', U.S. Department of State (17 August 1982) [https://www.ait.org.tw/declassified-cables-taiwan-arms-sales-six-assurances-1982/]
35 Alperstein, Ben, 'How Do Media Organizations Define Taiwan?', *The Diplomat* (6 November 2024) [https://thediplomat.com/2024/11/how-do-media-organizations-define-taiwan/]

ELEVEN — THE INDISPENSABLE ISLAND

1 'Taiwan's dominance of the chip industry makes it more important', *The Economist* (6 March 2023) [https://www.economist.com/special-report/2023/03/06/taiwans-dominance-of-the-chip-industry-makes-it-more-important]
2 Nienaber, Michael, 'Germany urges Taiwan to help ease auto chip shortage', Reuters (24 January 2021) [https://www.reuters.com/article/technology/germany-urges-taiwan-to-help-ease-auto-chip-shortage-idUSKBN29T04P/]
3 Lau, Stuart, 'Build better ties instead of only asking for microchips, Taiwan tells Europe', *Politico* (18 June 2023) [https://www.politico.eu/article/taiwan-better-relations-microchip-technology-tsmc-europe-germany-joseph-wu/]
4 Lau, 'Build better ties instead of only asking for microchips'
5 Doyle, Gerry, 'German ships that passed through Taiwan Strait were warned off, officials say', Reuters (2 October 2024) [https://www.

reuters.com/world/german-ships-that-passed-through-taiwan-strait-were-warned-off-officials-say-2024-10-02/]

6 Kowalcze, Kamil, 'TSMC Breaks Ground on €10 Billion German Plant in Chip War Salvo', Bloomberg News (20 August 2024) [https://www.bloomberg.com/news/articles/2024-08-20/tsmc-breaks-ground-on-10-billion-german-plant-in-chip-war-salvo]

7 Hawkins, Mackenzie, 'TSMC's Arizona Chip Production Yields Surpass Taiwan's in Win for US Push', Bloomberg News (24 October 2024) [https://www.bloomberg.com/news/articles/2024-10-24/tsmc-s-arizona-chip-production-yields-surpass-taiwan-s-a-win-for-us-push]

8 Hawkins, 'TSMC's Arizona Chip Production Yields Surpass Taiwan's in Win for US Push'

9 Sasaki, Reina, 'TSMC, Other Chipmakers Set to Make $80 Billion for Southern Japan' (6 September 2024) [https://www.bloomberg.com/news/articles/2024-09-06/tsmc-other-chipmakers-set-to-make-80b-for-southern-japan]

10 Koretski, Katherine, 'Vivek Ramaswamy defends positions on Ukraine and Taiwan', NBC News (30 August 2023) [https://www.nbcnews.com/meet-the-press/meetthepressblog/vivek-ramaswamy-defends-positions-ukraine-taiwan-rcna102408]

11 Zakaria, Fareed, 'On GPS: Is Taiwan's tech under threat?', CNN (31 July 2022) [https://edition.cnn.com/videos/tv/2022/07/31/exp-731-taiwan-tech-mark-liu-tsmc.cnn]

12 'Light and Lasers', ASML (2024) [https://www.asml.com/en/technology/lithography-principles/light-and-lasers]

13 'Q3 2024 Taiwan Semiconductor Manufacturing Co Ltd Earnings Call', TSMC (17 October 2024) [https://investor.tsmc.com/chinese/encrypt/files/encrypt_file/reports/2024-10/b474da862d1c24b1aa0c635a9f771261d93d3154/TSMC%203Q24%20Transcript.pdf]

14 Picker, Leslie, 'CNBC Exclusive: CNBC Excerpt: Citadel CEO Ken Griffin Speaking with CNBC's Leslie Picker on "Squawk on the Street" Today', CNBC (30 January 2024) [https://www.cnbc.com/2024/01/30/cnbc-exclusive-cnbc-excerpt-citadel-ceo-ken-griffin-speaks-with-cnbcs-leslie-picker-on-squawk-on-the-street-today.html]

15 'Oral History: Morris Chang', Computer History Museum (24 August 2007) [https://www.semi.org/en/Oral-History-Interview-Morris-Chang]

16 Miller, Chris, *Chip War: The Fight for the World's Most Critical Technology* (Scribner, 2022), p. 93
17 'Oral History: Morris Chang'
18 'Oral History: Morris Chang'
19 Mozur, Paul and Liu, John, 'The Chip Titan Whose Life's Work Is at the Center of a Tech Cold War', *The New York Times* (4 August 2023) [https://www.nytimes.com/2023/08/04/technology/the-chip-titan-whose-lifes-work-is-at-the-center-of-a-tech-cold-war.html]
20 'Morris Chang's book reveals he once asked Jensen Huang to lead TSMC', *Focus Taiwan* (8 November 2024) [https://focustaiwan.tw/business/202411080022]
21 'Taiwan "in middle" of AI revolution: Nvidia CEO', *Taipei Times* (25 March 2024) [https://www.taipeitimes.com/News/taiwan/archives/2024/03/25/2003815436]

EPILOGUE — EYE OF THE TYPHOON

1 Wong, Enoch, 'China confirms PLA Navy's new Type 076 amphibious assault ship will be drone carrier', *South China Morning Post* (13 January 2025)[https://www.scmp.com/news/china/military/article/3294568/china-confirms-pla-navys-new-type-076-amphibious-assault-ship-will-be-drone-carrier]
2 Sutton, H. I., 'China Suddenly Building Fleet Of Special Barges Suitable For Taiwan Landings', *Naval News* (10 January 2025) [https://www.navalnews.com/naval-news/2025/01/china-suddenly-building-fleet-of-special-barges-suitable-for-taiwan-landings/]
3 Shugart, Tom, X (20 January 2025) [https://x.com/tshugart3/status/1881429144912933126]
4 Mahadzir, Dzirhan, 'Japanese Fighter Scrambles Against Russian Threats on the Rise', *USNI News* (23 January 2025) [https://news.usni.org/2025/01/23/japanese-fighter-scrambles-against-russian-threats-on-the-rise]
5 Staff, 'Defending Taiwan "critical," says Rubio at US Senate hearing', *Focus Taiwan* (16 January 2025) [https://focustaiwan.tw/politics/202501160011]
6 Pamuk, Humeyra and Psaledakis, Daphne, 'US issues broad freeze on foreign aid after Trump orders review', Reuters (25 January 2025)

[https://www.reuters.com/world/us/trump-pause-applies-all-foreign-aid-israel-egypt-get-waiver-says-state-dept-memo-2025-01-24/]

7   Quinn, Jimmy, 'Rubio Exempts Taiwan and Philippines Security Programs from Aid Freeze', *National Review* (23 February 2025) [https://www.nationalreview.com/corner/rubio-exempts-taiwan-and-philippines-security-programs-from-aid-freeze/]

8   Shepardson, David and Holland, Steve, 'Trump and TSMC announce $100 billion plan to build five new US factories', Reuters (4 March 2025) [https://www.reuters.com/technology/tsmc-ceo-meet-with-trump-tout-investment-plans-2025-03-03/]

9   'Remarks by President Trump on Investment Announcement', The White House (3 March 2025) [https://www.whitehouse.gov/remarks/2025/03/remarks-by-president-trump-on-investment-announcement/]

10 'Press Note: USTBC President Offers Follow-up Comments Examining Data on Taiwan Arms Sales', US–Taiwan Business Council (20 September 2024) [https://www.us-taiwan.org/resources/ustbc-president-follow-up-comments-examining-data-on-taiwan-arms-sales/]

11 Chen Cheng-yu and Garcia, Sam, 'Legislative Yuan slashes government budget by 6.6 percent', *Taipei Times* (22 January 2025) [https://www.taipeitimes.com/News/front/archives/2025/01/22/2003830620]

12 Staff, 'Lai vows to lift defense spending to 3%', *Taipei Times* (15 February 2025) [https://www.taipeitimes.com/News/front/archives/2025/02/15/2003831908]

13 Chen, Kelvin, '1st batch of Taiwan's 1-year conscripts wrap up service', *Taiwan News* (26 January 2025) [https://taiwannews.com.tw/news/6023936]

14 Aspinwall, Nick, 'Taiwan's Indigenous to Xi: Taiwan Is Not China, Xi Does Not "Understand Dignity"', *The News Lens International* (9 January 2019) [https://international.thenewslens.com/article/111667]

FURTHER READING

Andrade, Tonio, *Lost Colony: The Untold Story of China's First Great Victory over the West* (New Jersey: Princeton University Press, 2011)

Chou, Catherine Lila and Harrison, Mark, *Revolutionary Taiwan: Making Nationhood in a Changing World Order* (New York: Cambria Press, 2024)

Cohen, Jerome A., *Eastward, Westward: A Life in Law* (New York: Columbia University Press, 2025)

Hioe, Brian, *Taipei at Daybreak* (London: Repeater Books, 2025)

Kerr, George H., *Formosa Betrayed* (Manchester: Camphor Press Ltd, 2018)

Lin, James, *In the Global Vanguard: Agrarian Development and the Making of Modern Taiwan* (California: University of California Press, 2025)

Miller, Chris, *Chip War: The Quest to Dominate the World's Most Critical Technology* (London: Simon & Schuster, 2022)

Nachman, Lev, *Contested Taiwan: Sovereignty, Social Movements, and Party Formation* (Washington: University of Washington Press, 2025)

Pottinger, Matt (ed.), *The Boiling Moat: Urgent Steps to Defend Taiwan* (California: Hoover Institution Press, 2024)

Rigger, Shelly, *The Tiger Leading the Dragon: How Taiwan Propelled China's Economic Rise* (Maryland: Rowman & Littlefield, 2021)

Ryan, Shawna Yang, *Green Island: A Novel* (London: Vintage, 2016)

Shirane, Seiji, *Imperial Gateway: Colonial Taiwan and Japan's Expansion in South China and Southeast Asia, 1895–1945* (New York: Cornell University Press, 2022)

Shirley Lin, Syaru, *Taiwan's China Dilemma: Contested Identities and Multiple Interests in Taiwan's Cross-Strait Economic Policy* (California: Stanford University Press, 2016)

Wei, Clarissa, *Made in Taiwan: Recipes and Stories from the Island Nation* (New York: Simon Element, 2023)

INDEX

27 Brigade (guerrilla fighters) 72
228 Incident 69–81, 84, 98, 104, 108, 138–9, 148, 153, 160, 184
228 Incident Settlement Committee 72, 73, 153
228 Incident Task Force 138–9
228 Memorial Park 80, 150, 160
1992 Consensus 161–2, 173, 178, 184, 185–6, 203, 208, 223
2758 Initiative 221
  see also Resolution 2758

Admiralty 274
Afghanistan 241
Africa 199
Age of Discovery 40
agriculture 31, 45, 46, 59, 183
  agricultural labour 33, 42, 45
  stockpiles 67
Air Defence Identification Zone (ADIZ) 191
Albania 216
Alishan Range 31
Allies 54, 58, 60–1, 116, 117, 152
  invasion of Normandy 193, 274
altars 24–5
Altmaier, Peter 256
American Chamber of Commerce (AmCham) 239
American Institute in Taiwan (AIT) 241, 243, 278
Amis people (Pangcah) 13–14, 17, 18, 20

Amnesty International 102, 121, 154
Amoco 121
ancestral spirits 15, 19–20, 78
Ando Rikichi 60
Andrade, Tonio 41
animism 15
Ankeng Execution Ground 91
Ann Arbor 155
anti-Communism 88, 95–6, 100–1
anti-KMT parties 93, 114–15
Anti-Media Monopoly protests 177–8
anti-sedition laws 109–10
apartheid 19
Apple 261, 270, 271
Aquino, Benigno (aka Noynoy) 130
Aquino, Corazon 130
Aristotle 196
Arizona 276
arms production 281
arms sales 128, 240–3, 278–9
Arrigo, Linda Gail 169
artificial intelligence (AI) 270–2
Asia Pacific Economic Cooperation (APEC) 183
Asian Tigers 30
assassinations 122–7, 129, 130, 131–3
Associated Press 127
Association for Relations Across the Taiwan Straits (ARATS) 162, 175
Atayal 14, 20
Australia 6, 9, 36, 110, 193, 198, 201, 221, 225, 262, 278

Index

Australian Financial Review (newspaper) 36, 202
Austria 189–90
Austronesian Expansion 23
automotive sector 255–6
aviation 55–7, 59–60, 198

Babuza 14
Baden-Wurttemberg (frigate) 257
Baguio 53
Bai-Yue lineage 36
Baidu 247
Baltic Way 166
Bamboo Union triad 122, 130–3
Bandung Conference 93
banyan trees 29, 74
baozi buns 91
baseball 58
Batavia (now Jakarta) 40
Beijing 5, 8–9, 21, 26, 64, 79, 93, 102, 107, 114, 118–19, 121, 137–8, 162, 164, 173, 185–6, 190, 196, 198–201, 205, 208, 210, 216, 218, 220, 222–3, 225–6, 228, 231, 233–5, 238, 241–5, 247–50, 256, 260, 263, 274–5, 277, 280
 claim on Taiwan 4, 6, 9, 11, 23, 35–6, 38, 282
Bell, Rear Admiral Henry H. 48–9
Bell Laboratories 264
benshengren (native Taiwanese) 88, 140, 145
Bera, Ami 277
Berlin 256, 257
BFMTV 201
bicycles 30
Biden, Joseph R. 182, 205, 240–1, 278–9
Blinken, Antony 233–4
Bloomberg 9
Bo Yang 101–2
boba 31, 35
Boeing 236–7, 262
Bosch, Robert 257
Boston 103
Britain 61
 see also United Kingdom
British Broadcasting Corporation (BBC), Mandarin Service 247
British Empire 47
British trading houses 32

Brown, William 128
bubonic plague 67–8
Buddhism 32
bullet train 28–32
Bunun 14
Bush, George H. W. 217, 278
Bush, George W. 279
Bush administration 164–5

Caijing magazine 244–6
Cairo Conference, 1943 60–1
Cairo Declaration 61, 100
Caixin Media 247–8
Cambodia 174
camphor 47
Canada 11, 19, 221, 225
Carter, Jimmy 119, 231, 236, 239, 240–1
Carter administration 118, 237
Cassidy, Rick 258
Catholics 15
CBS 230
censorship 85, 97, 101–2, 106–8, 148, 196
Central America 218
Central Asia 37
Central and Eastern European States 227
Central Intelligence Agency (CIA) 241
Central Mountain Range 17
Central News Agency 111
Chan I-hua 109–10
Chandler, Claire 221
Chang, Morris 264, 265, 267–72
Chang Yi-jung 86–7
Changhua 31
Chen Chen-hsiung 102
Chen Chi-li 132
Chen Chi-sen 158
Chen Chien-jen 15
Chen Chih-hsiung 92–4
Chen Di 37, 38
Chen Li-an 138
Chen, Mark 171
Chen Shui-bian (A-bian) 34, 110, 121, 134, 141, 156, 157–68, 170–6, 181, 184, 185
Chen, Vonny 93–4
Chen Wei-ting 179
Chen Wen-chen 125–7, 129

INDEX

Chen Yi (Communist) 102
Chen Yi (ROC Governor of Taiwan) 59–60, 61, 63–9, 72, 73, 75–6, 84, 91
Chen Yunlin 175–6
Cheng Chu-mei 110–11
Cheng Nan-jung (Nylon Cheng) 107–11
Cheng Nan-jung Liberty Foundation 111
Cheng Tzu-tsai 115
Chengdu 83
Chialung, Lin 10
Chiang Ching-kuo 83, 85, 90, 95–6, 98, 100–3, 114–20, 123–4, 127–9, 131, 133, 135–6, 140, 159–60, 169–70, 220, 232, 237, 266
Chiang Dynasty 234
    fall of 113–34, 222
Chiang Kai-shek 34, 54, 60–1, 63–4, 73–4, 78, 80, 113, 123, 140, 147–8, 153, 160, 203–4, 235–6, 276
    cult of 101
    death 113, 115–17, 133, 148, 169, 232, 236
    mausoleum 100
    retreat to Taiwan 83–6, 89, 92, 95–8, 100–3, 113, 153
    and the United Nations 216, 217, 219–20, 232
Chiang Kai-shek Memorial Hall 137, 176
Chiang Wan-an 80–1
Chiang Wei-kuo 123
Chiang Wei-shui 56
Chiangification 100
Chiayi 31, 72, 176
Chieh Shou Hall 54
Chien, Frederick 229–32
China 56, 74
    'century of humiliation' 196
    and Chiang Kai-shek 113
    and the Dutch East India Company 40
    early lack of interest in Taiwan 42
    economic dependence on 178, 179, 181, 207, 209
    and Japan 65
    and Koxinga 43–4
    New 99
    and the Second World War 59
    Taiwan's proximity to 38–9
    Three Kingdom Period 37
    and the United Nations Security Council 215
    see also People's Republic of China (PRC)
China Central Television (CCTV) 249, 273
China Coast Guard 191
China Times (newspaper) 158
Chinese Air Force 63
Chinese citizens, repression and fear endured by 5, 263
Chinese Civil War 169, 200
Chinese colonialism 14
Chinese Communist Party (CCP) 2, 3, 25, 102, 144, 173, 178, 207, 273
    and the 1992 Consensus 185
    claim on Taiwan 21, 35–6, 38, 77, 200, 202–3, 223, 252
    and the Falun Gong spiritual movement 197
    and Hong Kong 210–11
    and the Kuomintang 202–4, 213, 223, 265, 280
    and Lee Teng-hui 138, 139
    'mainland' 89
    and the media 247, 249
    military arm *see* People's Liberation Army
    and the possible invasion of Taiwan 9, 195, 197
    reverence for Xi Jinping 101
    and Su Beng 145
    Taiwanese resistance towards 5, 187
    and Taiwan's legislature 4
    target missiles against Taiwan 165
    and Tiananmen Square 199–200
    and the United Nations 216, 217–18, 223, 229
    and the White Terror 107
    willingness to impose suffering on the Chinese people 263
Chinese cultural identity 98, 99
Chinese culture 99, 145, 185
Chinese democracy activists 197–8
Chinese disinformation 4, 11
Chinese elites 46, 196
Chinese expansionism 10, 23, 185, 278, 282
Chinese folk religion 32
Chinese identity 4, 98, 99, 175, 209, 213
    imposition on Taiwan 18, 31

Chinese labourers 33, 42, 45
Chinese Monopoly Bureau 69–71
Chinese nationalism 61, 63, 77, 98–9,
    201–2, 249
Chinese Nationalist Party *see* Kuomintang
Chinese submarines 8
cholera 68
Chongqing (formerly Chungking) 204
Chou Ching 70
Chou, Tina 127
Christianity 15, 116
Christopher, Warren 229–32
Chu, Eric 212
Chung Lee-ho 71
Churchill, Winston 60, 116
cigarettes 69–70
Cihu, Taoyuan County 116, 133
Citadel 262
climate change 2, 193
Clinton, Bill 6, 138, 279
Clinton, Hilary 211, 248
Closer Economic Partnership Agreement 178
CNBC 262
coal 47
Coastal Mountain Range 17
Coca-Cola 236
Cochin (now Kochi) 39
Cohen, Jerome 117–19, 121–2, 129–32
Cold War 39, 124, 198, 222, 236, 282
colonialism *see* Chinese colonialism; Dutch
    colonialism; Japanese colonialism;
    Western colonialism
communicable diseases 67–8
Communism 96, 102, 117, 140
  see also anti-Communism
Communist revolution 65, 264
Communists 98, 123, 146, 168–9, 176, 204,
    207–8, 216, 237
  conquest of northern China 83
  conquest of southern China 84
  see also Chinese Communist Party
concentration camps 202
Confucianism 32, 45, 100, 145
  see also neo-Confucianism
Congress 124–5, 155, 163, 231, 238–41,
    249, 277

constitutional court 4, 104
Contras 241
Conway, Lynn 268
coral snorkelling 91
cost of living 1
courts 4, 104, 180
Covid-19 pandemic 7, 215, 229, 255–6
Coyett, Frederick 44
crime, organized 161
Cross-Strait Service Trade Agreement
    (CSSTA) 178
cultural genocide 201
Cultural Revolution 99, 222, 263
culture
  Chinese 99, 145, 185
  Taiwanese 185
cyberattacks, Chinese 4, 191
Czech Republic 225

Dadaocheng 69–70
Dahu 30
Daly City 122
Daxi 29
Defence Intelligence Bureau 132
democracy 115
  global 11
  recession in Asia 9, 263
  see also elections; Taiwan, democracy
'democratic Great Wall' 166–7
Democratic Progressive Party (DPP) 1–2, 85,
    133–5, 137–8, 141, 149–50, 155–6,
    158–61, 163, 165, 168, 170–1, 173, 175,
    177, 179, 181, 184–6, 204–6, 212–13,
    221, 227, 247, 275, 280
Den Kenjirō, Baron 57
Deng Xiaoping 137, 146, 236–8, 241, 265
Detention (2019) 106, 107
Díaz-Balart, Mario 277
Disneyland resorts 207
Dongfan 38
Dongfeng-26 (Guam Express) 6
Dongshe Township, Taitung County 91
Dongyin 192
Dresden, Germany 257, 259
drones 10, 273, 281
drought 2

INDEX 315

Dujarric, Stéphane 218–19
Durdin, Tillman 74
Dutch colonialism 14, 16, 22, 32–3, 39–45, 93, 186, 221
Dutch East India Company (VOC) 40, 41, 44

East Asia 39–40, 174
East China Sea 43, 278
East Pennsylvania State Penitentiary 51
East Rift Valley 17
East Turkestan (Xinjiang) 9–10, 198, 200–2, 218, 241
Easter island 23
economic dependence, on China 178, 179, 181, 207, 209
economic growth and development 25–6, 58–9, 65, 113, 163, 266–7
economic independence 178
economic mismanagement 65, 67
Economist, The (newspaper) 252
education 57, 59, 63, 65, 69, 79, 101, 199, 266
  Taiwan-centric 181
Egypt 60–3
elder care 1
elections 119–20, 136–7, 138, 143, 148, 149, 161, 205, 233
  1996 presidential 170
  2000 presidential 134, 140, 156, 161, 166–8, 170, 175, 185
  2004 presidential 134, 164, 167–8
  2008 presidential 134, 168, 173, 175
  2012 presidential 177
  2016 presidential 186, 203, 212
  2018 local 181, 205, 206, 209
  2020 presidential 181, 209, 210–12
  2024 presidential 216, 277
English Channel 193
entrepreneurialism 113
Estonia 166
Eurasian Plate 17
Europe 2, 3, 4, 11, 39–40, 139, 197, 257
European Union (EU) 166, 221, 258
Europeans 39
executions 9, 64, 74–5, 84–8, 91, 93–5, 98, 102, 146–7, 153
  letters 86–8

Executive Yuan 179
exports 183, 207
extradition 130–1
extreme ultraviolet lithography (EUV) 261

Facebook 250
Fahey, Michael 78–80, 172–3
Fairchild Semiconductors 261
Falun Gong spiritual movement 197
famine 107
Fan Yun 137–8
Fang Su-min 121
Far East/American Council of Commerce and Industry 114
fear, climate of 5, 98, 103, 146, 263
Feldman, Harvey 101, 103, 237–8
feminism 117–18
ferries 194–5
fertility rate 1
Filipino workers 6
film industry 106–7
financial crisis 2008 269–70
Financial Times (newspaper) 132–3, 192, 252
First Island Chain 8, 196
First Nations 19
First World War 61
fishing, Chinese 3
Forbidden City 200
Ford, Gerald 117
Ford Motor Company 236, 262
forestry 46
Formosa 39, 53, 57, 61, 62, 74, 95–6, 97
  see also Taiwan
Formosa Magazine 120–1, 246
Formosa TV 172
Formosa Yacht Resort, Anping 27
Formosan Association for Public Affairs (FAPA) 155
Formosan Expedition 48
Fort Santo Domingo 47
France 25, 49–50, 201, 215
Frankfurt am Main (logistic ship) 257
'Free China' ('Free Area of the Republic of China') 95–103, 114
Freedom Era Weekly magazine 108–9
Freedom of Expression Day 107–8, 111

'freedom of navigation' 257
Fu Hsing Kang College 89, 123
Fu Hsueh-tung 70
Fu Kun-chi 280
Fudekeng 78
Fuji, Mount 51
Fujian (aircraft carrier) 194
Fujian Province, China 33, 34, 42, 45, 64–5
Fujianese diaspora 33
Fujianese pirates 41, 43
Fung, Elmer 159

Garchik, Jerome 130
Gemeinschaft 140
General Agreement on Tariffs and Trade 182
General Motors (GM) 262
genocide 202
Germany 139, 189–90, 225, 256–9
'ghost money' 24
Ghost Month 24–5
Giant 30
GJ-II stealth drone 273
Glaser, Bonnie 224–5
Go Pek-hok (Ando Momofuku) 32
Goa 39, 40
God 15, 17
gold 98, 99
Graham, Katharine 133
Grand Hyatt Taipei 280
Grand Pacific Hotel, Suva, Fiji 35
Grand Shrine 54
graphics processing units (GPUs) 271
Great Hall of the People 21–2, 189, 200, 212
Great Leap Forward 107
Great Qing 33, 37
Green Island, Taitung County 90–1, 95, 102
Griffin, Ken 262
Gross Domestic Product (GDP) 178, 199, 266, 280
Guam 6
Guangdong 34, 42, 83, 200
Guangzhou 274
Guanmiao District, Tainan 159
guerrilla fighters 72, 146, 148
Guterres, António 218
Guyanmuchan 249–50

Haig, Alexander 241–3
haigui ('sea turtles') 245–6
Haishenwai 190
Haiti 227
Hakka 30, 32, 34, 42, 55, 140, 166
Han Chinese 10, 20, 36, 46–7, 107
Han Kuo-yu 149, 159–60, 181, 206–13
Han Ming 43
Hangzhou Bay 43
Haq, Farhan 225–6
Harper, Stephen 19
Hawaii 6, 23
He Bin 44
healthcare 57, 59, 67–8, 223–4
  universal 138, 199
Hebei 79
heijin ('black gold') 161
Hermosa española 41–2
Hill, Bruce 125–6
Hille, Kathrin 132–3
Hiraga Kyoko 147
Hirchy, Ira D. 75–6
Hirohito, Emperor 153
Hiroshima 152
Hitler, Adolf, *Mein Kampf* 189–90
Hla'alua 14, 21
Hoanya 14
Hokkien language 33, 44
Hoklo 32, 33, 34, 42
Holy See 226
home ownership 1
Hong Kong 10, 30, 61, 97, 102, 106–7, 123, 129, 142, 174–5, 178, 198, 201, 207, 210–11, 218, 230, 237, 249–50, 263
  democratic movement 9
  'one-country, two systems' deal 210–11, 245
Hou Yu-ih 109, 180
House Foreign Affairs Subcommittee 123–5, 127–8
House of Representatives (US) 222
Houston 163
Hsiao Bi-khim 213, 221
Hsieh, Frank 131–4, 138, 155, 173
Hsieh Hsueh-hung 72
Hsieh, Roger 92

INDEX

Hsieh Tao-lung 56
Hsieh Wen-ta 55–7, 58
Hsinchu 29, 30, 42, 114, 176, 207
Hsinchu Science Park 29, 266–7
Hsu Tzong-li 144
Hu Jintao 164
Hualien 17, 192
Huang Chao-chin 68
Huang Chun-lan 86–7
Huang Hua 137
Huang, Jensen 271, 271–2
Huang Kuo-chang 179
Huang, Peter 114–15
Huang Wen-kung 86–7
Huang, Wendell 262
Hudong shipyard, Shanghai 274
Hudus Haitang 14
human rights 9, 16, 23, 48, 56, 96, 107–8, 118–20, 125, 127, 133, 169, 174, 217–18
hungry ghosts 24–5
Hurley, Patrick J. 204
Huwei Airport 72
hyperinflation 265

identity *see* Chinese identity; Taiwanese identity
identity politics 47–8
illiteracy 59
immigration 1
  from China to Taiwan 10, 46–7, 83–4, 88, 98, 108
  Han Chinese 10, 46–7
  Japanese 57–8
  see also migration
Imperial Japan 2–3, 31, 55, 59, 61
  'pacification' campaigns 3
  see also Japanese colonialism
Imperial Japanese Army 50, 62, 92, 139, 204
Imperial Japanese Navy 50, 62
imports, foreign 183
India 39
Indian embassy (unofficial), Taipei 280
Indigenous Language Act 20
Indigenous people of Taiwan 13–23, 30, 32–4, 36, 38–9, 41–2, 46–9, 160, 166, 282
  and intermarriage 42

and Japanese colonialism 52–3, 57
and Koxinga 44
and the White Terror 88
and Zheng Jing 45
Indo-Pacific 40
Indonesia 23, 34, 92–3
Indonesian nationalism 93
industrial stockpiles 67
Industrial Technology Research Institute (ITRI) 266, 269
industry 45
  see also manufacturing
Infineon Technologies 257
inflation 265
infrastructure 58–9
Institute for National Defense and Security Research (INDSR) 194
Institute of Taiwan History, Academia Sinica 68
Intel 261, 269, 270, 271
Intelligence Section 90
Inter-Parliamentary Alliance on China (IPAC) 221
intermarriage 42
International Civil Aviation Organization 217
International Olympic Committee (IOC) 183
International Organization for Standardization (ISO) 227–8
internet services 281
iPhone 270
Iran 253
Itō Aviation Academy, Tokyo 55

Japan 6–10, 25, 28, 37–9, 47, 49, 79, 110, 114, 121, 147–8, 174, 198, 204, 236, 275, 278
  and a Chinese invasion of Taiwan 193, 262
  early lack of interest in Taiwan 42
  post-war 92
  and the Republic of China 181, 225, 227, 257
  and the Second World War 58, 61, 92–3, 151–3
  and the semiconductor industry 197, 257, 259, 265
  see also Imperial Japan
Japan Air Self-Defence Force 275

Index

Japanese colonialism 2–3, 16, 18, 22, 25, 29, 31–3, 37, 46, 48, 50–4, 55–61, 65, 72, 79–80, 90–1, 98–9, 114, 135, 139–40, 151, 161, 168, 176, 183–4
  administrative buildings 174
  end of 59–60, 62–3, 65–9, 216
  and the pacification of Taiwan 52
  sovereignty over Taiwan 65
  and Su Beng 145–6, 149
  and Taiwanese military conscription 53
  Taiwanese resistance to 52, 56–7
Japanese expansionism 49–50
Japanese migrants 57–8
Japanese suicide squads 62
Japanese traders 41
Japanese war criminals 73
Java 40
Jiangsu 123
Jiangxi province 237
Jin Canrong 196–9
Jingfu Gate 179
Jingmei Detention Centre 91, 92, 102, 121
Jinjiang Hotel, Shanghai 234
Joint Communiqué on the Establishment of Diplomatic Relations (Second Joint Communiqué), 1979 236–8
'Joint Sword – 2024B' (military drill) 191–2
Judicial Yuan 136

Kabayama Sukenori 50–2, 59
Kahabu 14
Kanakanavu 14, 21
Kanapathy, Ivan 277–8
Kangxi Emperor 45
Kaohsiung 42, 61–2, 73, 149, 153, 166, 168–9, 176, 192–3, 206–7, 211
Kaohsiung Incident, 1979 120–2, 126, 129, 134, 141, 148, 157–8, 168, 170
Kaolut 48–9
Kavalan 14
Keelung 24–5, 42, 50–1, 58, 61–2, 68, 73, 84, 147, 153, 166, 192–3
Keelung River 89
Kennedy, Edward M. 119
Kennedy, John F. 126
Kerr, George 53, 58, 63, 66–7, 73–4, 155

Ketagalan 14, 160
Kilby, Jack 264
Kim Dae-jung 130
Kinmen (previously Quemoy) 43, 163, 169, 192, 232
Kirk, Alexander Comstock 60
Kissinger, Henry 219, 235–7, 242
Ko Wen-je 180
Kolas Yotaka 17–20, 156
Koo clan 31
Korean War 197, 263
Koxinga (Zheng Chenggong) 16, 22, 33, 42, 43–4
Kumamoto, Kyushu 259
Kung family 64
Kuo, Michelle 25
Kuomintang (KMT) 18, 22, 30, 34, 36, 45, 48, 119–24, 126–7, 129, 131–3, 168–71, 173, 176, 178–81, 202, 205–6, 210–13, 222, 224, 227, 231, 252, 264–6, 273, 275
  and the 1992 Consensus 185
  and the '228 Incident' 69–81
  anti-KMT organizations 93, 114–15
  armed uprisings against 94–5
  and Chen Shui-bian 159–63, 165, 167, 168
  and Chen Yi's tenure 65–9, 72, 75–6
  and the Chinese Communist Party 202–4, 213, 223, 265, 280
  claim on Taiwan 3, 200
  and the death of Chiang Kai-shek 116
  defence spending 279, 280
  and the Democratic Progressive Party 133–4
  end of their monopoly on political power 246
  and 'Free China' 96–9, 101–2
  and IPAC 221
  and the Kaohsiung Incident 158
  and Lee Teng-hui 136, 140–1, 143, 144
  as minority party 216
  and Peng Ming-min 150, 153, 154, 155
  as protectors of Chinese culture 99
  retreat to Taiwan 54, 59–60, 62–3, 83–4, 200, 202
  return to power, 2008 175
  and the semiconductor industry 259–60

Index

and Su Beng 146–9
Taiwan Garrison Command 73
and Taiwanese independence 108–10, 135
and the White Terror 85, 87–98, 102–7, 111, 169–70, 184, 209
Kyoto 93
Kyushu Financial Group 259

labour 140
 Chinese agricultural 33, 42, 45
 forced 91
 shortages 1
Lai Ching-te 1, 4–5, 10, 11, 27–8, 32–5, 36, 144, 175, 185–6, 190–2, 213, 221, 226, 233–4, 273, 277, 279–81
Lai Pin-yu 179
Lan Po-shih 145
land concessions 47
land reforms 113, 146
Lantos, Tom 128
Latvia 166
Leach, Jim 127–8
Lee Teng-hui 34, 135–44, 155, 161, 162, 166–7, 170, 182, 183, 184, 222
Legislative Yuan 119, 148, 159, 163, 168, 179, 186, 211–13, 275, 279, 280
Lewis, Maggie 104
Li, K. T. 263, 264–5, 265–7
Liao, Thomas 93
Liaoning (aircraft carrier) 192
Liberty Square 176–7
Lieh, Lee 104–7
Lien Chan 138, 141, 161, 165, 167
Lilley, James 243
Lin Chian-mai 69–70
Lin Fei-fan 174–5, 177–81
Lin, James 46–7, 59, 220
Lin Join-sane 178
Lin Yang-kang 138
Lin Yi-hsiung 120–1
Lin family of Wufeng 46
Lithuania 166, 225
Liu Chieh 219–20
Liu, Helen 122–4, 131–2
Liu, Henry (Chiang Nan) 122–33, 155
Liu Jieyi 207–8

Liu, Mark 197, 260
Liu Po-yen 176–7
Liu Xiaobo 108
Liu Yao-ting 87
Long March 237
Los Angeles 163
Lu, Annette 117–22, 126, 129–30, 134, 141, 161, 165, 167–8
Lu Shaye 201–2
Lukang 31
Luo Wen-jia 162
Luzon 37, 53
lychees 45

Ma Ying-jeou 103, 118, 129–30, 133–4, 161, 168, 173–5, 177–8, 180–1, 202–5, 208, 213, 223, 225, 245, 281
MacArthur, General Douglas 53
Macau 39, 40, 207
McCarthy, Kevin 192
Machangding, Taipei 74, 84, 91
Mackay, George Leslie 47
MacKenzie, Lt Cdr Alexander 48–9
Madagascar 23
'mainland', retaking of the 83, 89, 138, 154, 216, 222, 232, 282
Mainland Affairs Council 184
Makatao 14
'Make Taiwan China Again' 173
Malacca 39
Malala Yousafzai 108
malaria 44
Malaysia 23
Manchukuo 183
Manchuria 10, 61, 183, 190–1
Manchus 37, 42–3, 45, 47, 49–50, 61
 see also Qing Dynasty
Mandela, Nelson 19, 108
mangoes 45
Manila 6, 39, 53
manufacturing 140, 209
 see also industry
Mao Zedong 8, 33, 65, 77, 83, 96, 99, 107, 113–14, 117, 123, 143, 146, 197, 200, 203–4, 216, 222, 234, 244, 246, 249, 264, 265

martial law, Chinese 9, 18, 48, 76, 78, 80, 84–111, 115, 117–18, 123–4, 125, 128, 135, 140, 150, 154, 169–70, 231
  arrest process 90–2
  end of 85, 88, 92, 103–11, 137, 149, 170, 174, 209
  legacy 160
  yidu ('residual poison') of 89
  see also White Terror
Martial Law Section Jail 90
martyrs 109–11, 177
Marxism 145, 146
Matsu 163, 192, 232, 281
Matsu archipelago 3
Maxwell, James Laidlaw 47
Mead, Carver 268
media 246
  American 251–3
  Chinese 244–51
  control 97, 101–2, 106–7, 177–9
  freedom 110–11, 199
  pro-China 208–10
MediaTek 269
medical doctors 47
Mena House Hotel 60
Merkel, Angela 256
Miaoli, Kingdom of 30
Miaoli City 30
Miaoli County 30, 42, 166
Michigan Daily, The (student newspaper) 125–6
Microsoft 271
migration 23, 47
  mass Chinese 10, 33, 34, 42
  prehistoric 36
  see also immigration
Mineta, Norman 128
Ming Dynasty 33, 37–9, 41–6, 97, 200
Ministry of National Defense 97, 122
missionaries 47
Mohammed, Amina 225
Monument to the People's Heroes 200
Moore, Gordon 261
Moore's law 261–2
Motorola 269
Mujahideen 241
Murdoch, Rupert 208

Murray, Yvonne 218–19
Musha Village 53
Musk, Elon 276
Mutual Defense Treaty 231, 239

Nagasaki 39, 151–3
Nagorski, Andrew 230–1
Nanjing (formerly Nanking) 2, 43–4, 69, 74, 204, 265
National Affairs Conference, 1990 143
National Assembly 136, 137, 143
National Chengchi University 115
National Palace Museum 31, 141
National Press Club of Australia 201–2
National Review (magazine) 225
National Security Council 175, 278
National Taiwan University 125–6, 131, 137, 153, 154, 177, 244
nationalism
  Chinese 61, 63, 77, 98–9, 201–2, 249
  Indonesian 93
  Taiwanese 56
Nauru 226–7
Necessary State Socialism 64
neo-Confucianism 97
nephrite (Taiwanese jade) 23
Netherlands 40, 221
  see also Dutch colonialism
New Life Correction Centre 91
New Life movement 100
New Party 167
New Taipei City 29, 144
New York 114, 154, 163, 215, 218, 228, 247, 252
New York Police Department (NYPD) 114–15
New York Times, The (newspaper) 52, 74, 115, 136, 154, 251
New Zealand 23
Newsweek (magazine) 133, 230
Nicaragua 241
Ningxia night market 90
Nixon, Richard 113, 217, 219, 234, 236, 244
noodles, instant 32
Normandy, Allied invasion 193, 274
North America 2

INDEX

North Atlantic Treaty Organization (NATO) 7, 166
North Korea 7, 9, 164, 253
nuclear power 2, 163, 164
nuclear waste, storage 16–17
nuclear weapons 6, 7, 9, 151–3, 278
Nvidia 261, 271–2
NXP Semiconductors 257
Nylon Cheng Memorial Museum 107–8, 110

Oasis Villa prison 95
Obama, Barack 182, 205, 279
Office of the Governor-General 51, 54
Office of Republic of China Affairs 237
Okinawa 152
Olympic Games 45
  Summer Olympics, Beijing, 2008 244, 245
Omaha Beach 193
'One China Principle' (Beijing) 185, 222–3, 226, 228, 235, 260
one-China policy (Washington) 199, 203, 219, 222–3, 226, 244, 260
Ong Thiam-teng 66, 71, 72, 73
Opium Wars 61
  First 47, 196
Orchid Island 15, 16, 17
organized crime 161
Outer Manchuria 190–1

Pacific 7–8, 39, 196
Pacific Island Nations 23, 58, 61, 226
Paiwan 13–14, 20, 49
Palace Museum, Beijing 98
Palace Museum, Taipei 98–9
Pan Hsiao-hsia 109
Papora 14
Paradise Culture Associates 148
Paraguay 227
Parker, Robert 239–40
party votes 212
Paul, Weiss 130–1
Pax Japonica 59
Pazeh 14
peaches 45
peasantry 59
Pegasus Teahouse, Dadaocheng 69, 78

Pelosi, Nancy 192, 201
Peng Meng-chi 73, 153
Peng Ming-min 92, 135, 138, 149–56, 170
Penghu 25, 37, 40–1
People First Party 165, 167
People's Daily, The (newspaper) 249
People's Liberation Air Force 191
People's Liberation Army (PLA) 5–6, 185, 213, 263
  and an invasion of Taiwan 143, 193, 195
  and the Chinese Communist Party 143
  military drills around Taiwan 5, 191–2
  Rocket Force 6, 191
  status 197
People's Liberation Navy 191, 193–4, 273, 274
People's Political Council Assembly 68
People's Republic of China (PRC) 30, 77, 144, 154
  and Chen Shui-bian 161
  chequebook diplomacy 199
  claim on Taiwan 2, 3–9, 11, 35–8, 61, 99, 149, 189–92, 195, 199–203, 210, 223–6, 235–8, 244, 247, 250–3, 282
  and the 1992 Consensus 161–2
  general Chinese support for 195, 196
  illegality of 37–8, 202
  and the Indigenous people of Taiwan 23
  coastal economy 198
  coastline 7–8
  coercion and bullying of 278
  cognitive warfare on the Taiwanese people 4
  and Covid-19 229, 255
  crackdown on democracy protesters, 1989 263
  daily harassment of Taiwan 275
  economic slowdown 26, 263
  foundation, 1949 83, 176, 190, 191, 200, 215–16
  hatred of the Taiwanese people 202
  international image 199
  and Kaohsiung 207
  and Lee Teng-hui's presidency 138, 139–44
  and the media 244–51
  military build-up against Taiwan 173, 218
  military drills around Taiwan 5, 191–2, 245

People's Republic of China (PRC) – *cont.*
  and North Korea 7
  nuclear arsenal 6
  political stability 197–8
  and the possible invasion of Taiwan 7–8, 22, 163, 191–9, 210, 212–13, 222, 227, 260, 262–3, 273–5, 278, 281–2
    benefits of 196–9
    catastrophic nature 201
    moral and propaganda boost 8–9
    and the strategic landscape 7–8, 196
  re-education policy 201–2
  and the Republic of China 163–7, 170, 173, 175, 178, 180, 182, 184–6, 190, 199, 202–3, 207–10, 212–13, 216, 224, 280, 282
  and Russia 7, 275
  and the semiconductor industry 10, 258, 263
  and the severing of undersea cables 3–4, 281
  sovereignty over Taiwan 221
  and Su Beng 146
  target missiles against Taiwan 164, 165, 166–7
  and the theft of Taiwan's resources 3
  and the Tiananmen Square massacre 137
  as Trojan horse of a trade partner 143
  and the United Nations 216–20, 223–4, 228, 229
  and the United States 114, 119, 236
  see also China
Pescadores (now Penghu) 25, 40–2, 44, 50, 61
Philippine Sea 15
Philippine Sea Plate 17
Philippines 6, 8, 23, 34, 37, 39, 53, 130, 193, 262, 276
Phoenix, Arizona 258–60
Pingpu plains peoples 14, 16
Pingtung 42, 75–6, 92, 182
Plaza Hotel, Manhattan 114–15
plums 45
political officers 89
Political Work Department 123
Popeye (cartoon strip) 101–2
pornography 198
Portugal 39–40, 41

post-Cold War security order 10, 278
Potsdam Declaration 61
poverty 4–5, 20, 185
power grid 2
Presbyterian Church 15, 47–8
prisons 90–1, 94–5, 97, 121, 169–70
Provincial Police Headquarters Detention Centre 90
Provintia Castle 44
Putin, Vladimir 275
Puyuma 14
Pyongyang 164, 253

Qing Dynasty 16, 22, 25, 29, 32, 34, 42, 43–7, 49–50, 52, 61, 190, 200
  and Koxinga 43–4
  sovereignty 46
  Taipei city wall 179
  see also Manchus
Qingdao 146
Quinn, Jimmy 225

Radio Free Asia 250
Ramaswamy, Vivek 260
RCA 267
Reagan, Ronald 232, 241, 242–3
Reagan administration 124
Red Army 146
Red Guards 99
redshirt protests 171, 172
referenda 164–5, 167, 168
refugees, from China to Taiwan 83–4, 88, 98, 108
Reporters Without Borders 110
Republic of China (ROC) 2–3, 21, 22, 27–8, 35, 57, 101, 124
  arrival in Taiwan 54, 60, 61–3
  assimilation policies 16
  and the Cairo Declaration 100
  and Central America 218
  and Chen Shui-bian 161
  claim on Taiwan 61
  contemporary state 216
  and the death of Chiang Kai-shek 115–16
  and democracy 115
  and economic mismanagement 65, 67

and feminism 117–18
flag 252
foundation, 1911 99, 190, 191
and the governance of Chen Yi 63–9
and Japan 139
and the Kaohsiung Incident 158
and Lee Teng-hui's presidency 135–40
National Day 179, 190, 191
national flower 45
Office of the President 51
overthrown 108, 216
and Peng Ming-min 150, 153
and the People's Republic of China 163–7, 170, 173, 175, 178, 180, 182, 184–6, 190, 199, 202–3, 207–10, 212–13, 216, 224, 280, 282
retreat to Taiwan, 1949 54, 60–3, 83–4, 99, 200
as rump state 95–6
and the Second World War 58
and Su Beng 148, 149
support peaceful disobedience 180
and the United Nations Charter 216
and the United States 154, 275–8
Washington's derecognition of, 1979 113, 118, 148, 229–32, 239–40, 265–6, 275
and the White Terror 81, 83–111, 86, 89, 93, 98, 103, 105–6, 184
Republic of China Air Force 57, 63, 198
Republic of China Army 53, 62
Republic of China High Court 172
Republic of China Military 193, 231
mandatory service 114
Republic of China state radio 123
Republic of Formosa 50, 56
Republic of Taiwan 93, 109
Republic of Taiwan government-in-exile 93
Resolution 2758 216–17, 220–2, 225–7
rice 31, 45
ponlai 57–8
robot dogs 273–4
Rockefeller, Nelson 116–17
Rogers, William 114
Roosevelt, Franklin D. 60–1, 100, 116, 276
Rosenberger, Laura 226–7
Rover (merchant ship) 48, 49

Roy, J. Stapleton 237–8
RTÉ News 218
Rubio, Marco 275–6
Rudd, Kevin 19
Rukai 14
rule of law 63, 66
Russia 7, 9, 190–1, 195–6, 215, 223, 280
invasion of Ukraine 191, 195, 275, 276, 277
see also Soviet Union
Russian Far East 190–1
Ryukyu Islands (Okinawa Prefecture) 37, 49

Saint Lucia 227
Saisiyat 14
Sakizaya 14
same-sex marriage 174, 181, 199
Samsung 270, 271
San Francisco 123
treaty of 65
sand 3
sanitation 67
SARS 229
Satsuma Rebellion 259
Scholz, Olaf 257
School of International Relations 196
Schwarzenegger, Arnold 207
Sea of Japan 8
Second World War 53–4, 58–61, 79, 117, 129, 151–3, 168–9, 193, 215–16
secret police 85, 90, 96, 98
Seediq 14, 52–3
semiconductor sector 2, 10, 25, 29–30, 197, 255–70, 272, 276–7
Senate 155, 222, 239, 240–1, 275
Seoul 252
Separate Customs Territory of Taiwan, Penghu, Kinmen and Matsu, The 182–3
September 11 attacks 163–4
service sector 178
Shandong 245
Shanghai 60, 64, 98, 124, 178, 185, 200, 207
Shanghai Communiqué (First Joint Communiqué), 1972 234–5
Shangri-la Hotel, Singapore 202
Shapiro, Don 230
Shen Chang-huan 231

Shen Cong 137
Shen Ming-shih 194
Shen Yourong 38
Shenzhen 207–8
Shepherd, Mark 264, 265
Sheraton Grand Hotel 90
Shi Lang 45
Shih Ming-teh 158, 168–73
Shimonoseki, treaty of, 1895 50, 61
Shinto 54
Shugart, Tom 274–5
Shultz, George 243
Sichuan (Type 076 amphibious assault carrier) 273–4
Silicon Valley 265
Sina 247–8
Sinevaudjan 49
Singapore 23, 30, 202, 252
Sinification 45, 99–100
Sino–Japanese Wars
  First 50
  Second 113
Sinosphere 253
Siraya 14, 38
Six Assurances 232–4, 243, 244
Smith, Art 55
social media 249–50
social reform 181
Solarz, Stephen 124–5, 127
Soong, James 140–1, 161, 165, 167
Soong Mei-ling (Madame Chiang Kai-shek) 60, 64, 83, 84, 96, 115, 123
Soong family 64
South Africa 1, 19
South China Sea 8, 23, 205, 245, 277, 278
South Korea 7, 8, 9, 30, 110, 130, 193, 278
Southeast Asia 8, 23, 50, 53, 58, 65, 93, 174, 278
Southeast Asians 33, 34
Southern Ming 43
Southern Mongolia 10
Soviet Union 166, 241, 266
  see also Russia
Spain 39–40, 41–2
squid 3
Stalin, Joseph 116
Stinger missiles 278, 281

Straits Exchange Foundation (SEF) 162
student spies 102–3, 126, 127
Su Beng (Lîm Tiâu-hui) 135, 144–9, 155, 180, 186
Su Chi 161–2, 184, 185
Su Ching-hsuan 78
sugar cane 45
Sukarno 93
sulphur 47
Sun Yat-sen 100
Sunflower Movement, 2014 106, 174, 178–81
superaged societies 1
surfing 91
Syapen Nganaen 15, 16, 17, 20
Syria 88

Tagawa Matsu 43
Taichung (Taichū under Japanese rule) 30–1, 35, 55, 72, 176, 192, 193, 207, 244
Taihoku New Park (now 228 Peace Memorial Park) 80, 150, 160
Taihoku Prison 51
Taiko (now Dajia) 151
Tainan (formerly Taiwan-fu) 27–8, 32–3, 38, 43–5, 47, 50, 57, 68, 157, 159, 167, 170, 174, 176, 179, 271, 272
Taipei 3, 5, 7, 11, 13, 27–9, 31–2, 51, 53, 56, 58, 61–2, 65, 68–70, 73–4, 78, 80, 83–4, 90–1, 97–100, 104, 108, 110, 114, 116–18, 121, 124, 126–7, 129, 131, 137, 141, 145, 153, 157, 159–61, 163, 174–6, 179, 182, 186, 192–4, 206–7, 216, 221, 225–6, 229–32, 239, 242–4, 247, 252, 256, 263, 280
Taipei (aircraft) 56
Taipei Times (newspaper) 94
Taishang (Taiwanese business people with enterprises in China) 163
Taitung 17, 192
Taivoan 14
Taiwan
  defence 8, 10, 278–81
  democracy 3, 8, 25, 48, 57, 68, 77–8, 89, 92, 94, 102, 110, 119, 133–4, 136–8, 140, 143, 148, 150–1, 154–6, 157–87,

INDEX

191, 204–6, 211, 216, 222, 229, 246–7, 251, 280, 282
earthquakes 262
economy 25, 58–9, 65, 113, 163, 266–7
entry to the historical record 38–42
global presence 25–6, 30, 163, 182–3, 186
Gross Domestic Product 178, 199, 266, 280
and Koxinga 42, 43, 44
lack of international understanding regarding 11
liberal democracy's ghosting of 25–6
manufacturing 266
multiculturalism 32–3
as offshoot of 'mainland' China 89
place renaming 99–100
population 59, 192
and the Qing Dynasty 43–4, 45–7, 49–50
Republic of China's plundering of 65–7
Republic of China's retreat to, 1949 54, 60–3, 83–4, 99, 200
'retake the mainland' aspiration 83, 89, 138, 154, 216, 222, 232, 282
and the Second World War 58
semiconductor sector 2, 10, 25, 29–30, 197, 255–70, 272, 276–7
strategic location 53
tech emergence 265–7
Trump's attitude towards 10
unification 3
and the United Nations 215–21, 223–9, 232, 234
and the United States 8, 139, 163
uprisings against the Qing 46–7
and the World Trade Organization 182–3
and the world's balance of power 11
see also Formosa; 'Free China'; Japanese colonialism; martial law, Chinese; Republic of China; Republic of Formosa; Republic of Taiwan
Taiwan Affairs Office, Beijing 207
Taiwan Air Force 275
Taiwan Association for Truth and Reconciliation 78
Taiwan Beer 58
Taiwan Cultural Association 56
Taiwan Daily News (online news publication) 123

Taiwan Democratic Independence Party 93
Taiwan Exhibition 64–5
Taiwan Garrison Command 97, 126
Taiwan International Solidarity Act 222
Taiwan Miracle 113
Taiwan Pavilion 200
Taiwan People's Party (TPP) 180, 221, 279
'Taiwan problem' 198
'Taiwan question' 22, 222, 235
Taiwan Relations Act, 1979 233, 234, 240–1, 243, 244
Taiwan Representative Council 66
Taiwan Semiconductor Manufacturing Corporation (TSMC) 2, 29, 197, 255–63, 268–70, 272, 276–7
Taiwan Shin Sheng Daily News (newspaper) 139
Taiwan Stock Exchange 163, 191
Taiwan Strait 25, 28, 37, 44, 45, 61, 64, 98, 154, 163, 165–6, 193, 222, 223, 235, 243, 257, 277
Taiwan Travel Act 155
Taiwan-fu (later Tainan) 47
Taiwanese assembly 56
Taiwanese civil society 174, 176, 181
Taiwanese Communist Party 72
Taiwanese cuisine 32–5
Taiwanese culture 185
Taiwanese diasporic resistance movement 9–10, 92, 148, 155
Taiwanese elites 72, 76–7
Taiwanese emigrants 103, 148, 228–9
Taiwanese identity 3, 9, 28, 77, 88, 89, 99–100, 138, 139, 140, 148, 149, 154, 167, 175–6, 179, 204, 208–9, 213, 228
Taiwanese independence 88, 93, 94, 108, 114, 118–21, 135, 144, 146–51, 154–5, 164, 166, 170, 190
Taiwanese Independence Association 148
Taiwanese language 33–4
Taiwanese military 173
and conscription 53, 65, 114, 173, 281
nationalization 143
Taiwanese nationalism 56
Taiwanese resistance 9–10, 52, 56–7, 92, 148, 155, 187

Taiwanese self-determination 57, 88, 119, 135, 150–1, 155–6, 178, 179–80, 223
Taiwanese separatists 22, 162, 190
Taiwanese sovereignty 3, 5, 11, 30, 35–6, 143, 144, 156, 163, 173, 178, 180, 191, 210, 215–53, 273
Taiwanese state, establishment 72, 108, 149, 163, 203
Taiwanese vigilante units 67
Taiyuan Prison 91, 94–5
Takao (later Kaohsiung) 47, 58
Tamsui 47, 50
Tân Chhoàn-tē 72
Tang Fei, General 162–3
Tang, Jane 244–51
Tangshan (China) 89
Tangwai movement 119–20, 133, 135, 169–71, 174, 246
Tao 14, 15, 16, 17, 20
Taoism 32
Taokas 14
Taoyuan 29, 100
Taoyuan International Airport 29
Tapani Incident 1915 52, 57
tariffs 205
taxation 47, 59
tea 47
Tehran 253
telecoms fraud 198
Teng, Emma 45–6
Tesla 262
Texas Instruments 263–4, 265, 267, 269
Thailand 34, 174, 186
Thao 14
Third Joint Communiqué, 1982 242–3
Third Taiwan Strait Crisis 139
thought police 89, 123
three Joint Communiqués 233–4, 234–5, 236–8, 242–4
Tiananmen Square 83, 189, 199–200
  protests and massacre, 1989 77, 137, 195
Tianqi Emperor 41
Tibet (Xizang) 9, 10, 198, 200, 201, 218
Tientsin, treaty of 47
timber 31
Time magazine 95–7

To, Chapman 106
Tokyo 3, 50–1, 54–9, 65, 79, 145, 147–8, 151, 252, 279
Tomb Sweeping Festival 78
torture 9, 20, 64, 85, 90–2, 102, 120, 169
tourism 97, 198
trade 8, 32, 41, 143, 178, 238–9
trade war 205
transistors 261, 264
treaty ports 47, 61
Truku 14
Truman administration 204
Trump, Donald 10, 155, 182, 204–5, 211, 247–8, 258–9, 275–9
Trump administration 258, 260, 276, 277
Truth and Reconciliation Commission 19
Tsai Eng-Meng 208–10
Tsai Ing-wen 14, 16–17, 19–21, 34, 36, 110, 144–5, 149, 173, 177, 181–7, 192, 204–6, 208, 209, 211–12, 218–19, 224, 258, 265, 281
Tsai, Jewel 34
Tsai, Rick 270
Tsao Shan (Yangming Mountain/Grass Mountain) 84, 96–7
TSMC *see* Taiwan Semiconductor Manufacturing Corporation
Tsou 14
Tun-Hou Lee 103
Tungning ('Eastern Pacification') Dynasty 44–5
Tungtu ('Eastern Capital') Dynasty 44
Tutu, Desmond 19
Twitter (X) 250, 274
Typhoon Kon-rey 193
typhoons 193

Ukraine, Russian invasion of 191, 195, 275, 276, 277
Umbrella Revolution 106
undersea cables, severing 3–4, 281
Unger, Leonard 229
United Daily News (newspaper) 139, 158
United Kingdom 28, 60, 210, 215, 225, 227, 245
  see also Britain

INDEX

United Microelectronics Corporation (UMC) 267
United Nations (UN) 25, 37, 72, 75–6, 113, 153, 154, 215–29, 232, 234
United Nations Charter 58, 216, 217, 218
United Nations Educational, Scientific and Cultural Organization (UNESCO) 217
United Nations General Assembly (UNGA) 215, 216–19, 220, 225, 226
  Resolution 2758 216–17, 220–2, 225–7
United Nations Headquarters, Manhattan 215, 218
United Nations High Commissioner for Human Rights (OHCHR) 223
United Nations Security Council (UNSC) 76, 113, 215–17
United Nations Sustainable Development Goals (UNSDGs) 225
United States 4–10, 22, 25, 28, 47, 72, 92, 99, 102–3, 113–15, 117–18, 122, 154–5, 196, 222–3, 257, 271, 273
  and the '228 Incident' 76
  agricultural exports 183
  and an end to the US-led post-WWII order 278
  anti-communism of 95–6, 101
  and China 113–14, 117–19, 163–4, 198–9, 202, 204–5, 216, 226, 229, 233–43
  and China's possible invasion of Taiwan 193, 197, 203, 213, 240, 260, 262, 276–8
  defence aid to Taiwan 276
  derecognition of the ROC, 1979 113, 118, 148, 229–32, 239–40, 265–6, 275
  elections, 2016 247–8
  and the Kuomintang 63, 69, 122–3
  and media 251–3
  and the Republic of China/Taiwan 4, 8, 48, 60–3, 113–14, 116–18, 120, 122–33, 139, 154–5, 163–5, 181–2, 192, 197, 204–5, 222, 224–7, 229–44, 257–60, 262, 265–6, 275–8
  and the Second World War 53, 54, 58, 151–3, 168
  and the semiconductor industry 258–60
  sovereignty 128
  strategic ambiguity policy 260

Taiwanese emigrants 148, 228–9
tech sector 265
trade with Taiwan 238–9
and the United Nations Security Council 215
United States Department of State 65, 76, 128, 204, 231, 235, 237
United States Navy
  Asiatic Squadron 48
  Seventh Fleet 5–6, 138, 173
Uongu Yatauyongana 88
US–Taiwan Business Council 278
Uyghur Muslims 9, 164, 202

van Linschoten, Jan Huygen 40
Vance, JD 276
Vietnam 25, 34, 174, 195–7
Vietnam War 123
Volt Typhoon 6
von der Leyen, Ursula 257

waishengren (mainlanders) 83–4, 88, 98, 108, 140
Wałęsa, Lech 170
Wall Street Journal, The (newspaper) 251
Wall Street Journal Asia, The (newspaper) 122
Wang Hsi-ling, Admiral 130, 132
Wang Huning 280
Wang Jin-pyng 179–80
Wang Jingwei 64
Wang Mei-hua 256
Wang Sheng, General 123–4
Wang Ting-yu 280
Wang Yang-ming 96–7
Want Want China Times Media Group 208–9
War on Terror 164, 165
Warren Commission 126
Washington 7, 48, 77, 96, 113–14, 117–18, 123–4, 129–30, 150, 164, 204–5, 216, 220, 224–5, 232, 236–7, 239, 241–2, 244, 247, 252–3, 260, 265–6, 278–9
Washington Post, The (newspaper) 133, 251
Watergate scandal 236
wax apples 45
WeChat 249
Wecht, Cyril 126–7

Wei, C. C. 257, 258, 276
Weibo microblogging platform 248
Wen Jiabao 164–5
Weng Hsiao-ling 213
Western colonialism 14–15
White House 10, 114, 124, 164–5, 231, 236, 241, 247, 275, 276, 278
White Terror 81, 83–111, 184
  see also martial law, Chinese
Wild Lily Movement 137, 143, 176
Wild Strawberry Movement 176–7
Wokou 38
Wolff, Lester 102
Wong, Joshua 174, 249–50
World Health Assembly (WHA) 223–5
World Health Organization (WHO) 217, 223, 225, 229
World Trade Organization (WTO) 182–3
World United Formosans for Independence (WUFI) 114
Wu, Albert 25
Wu, Joseph 256–7
Wu, K. C. 124
Wu Shu-chen 157, 159, 168, 171–2
Wu Sz-huai 212–13
Wuhan 64

X (Twitter) 250, 274
Xi Jinping 3, 5, 7–8, 21–3, 35–6, 101, 149, 162, 182, 189–90, 195–7, 199, 201–3, 205, 210–13, 247, 249–50, 262–3, 274–6, 280–2
  'new era of socialism with Chinese characteristics' 196–7

Xi-Ma summit 2015 202–3, 204
Xiamen 44
Xiao Qian 36, 37, 201–2
Ximending 90
Xindian River 74, 91
Xinhua News (newspaper) 249
Xinjiang *see* East Turkestan
Xinxin magazine 66
Xizang *see* Tibet

Yangtze River 43, 64
Yeh Chu-lan 110–11
Yen Chia-kan 116
Yeung, Johnson 174–5
Yey Te-Ken 70
Yilan County 93
youth, disillusionment of 1
Yu Ching-fang 52
Yuli 17
Yunlin 31, 72
Yushan (Patungkuonu/Niitaka-san/Jade Mountain) 51

Zeelandia Castle 41, 44
Zelenskyy, Volodymyr 276
Zhao, Mrs 79–80
Zhejiang Province 34, 84, 97, 264
Zheng Jing 44–5
Zheng troops 45–6
Zheng Zhilong 43, 44
Zhongzheng District, Taipei 174
Zhou Enlai 146, 222, 235–6
Zhoushan 43
Ziede, James 114